cSUR-UT Series: Library for Sustainable Urban Regeneration
Volume 4

Series Editor: Shinichiro Ohgaki, Tokyo, Japan

cSUR-UT Series: Library for Sustainable Urban Regeneration

By the process of urban development in the 20th century, characterized by suburban expansion and urban redevelopment, many huge and sophisticated complexes of urban structures have been erected in developed countries. However, with conventional technologies focused on the construction of structures, it has become difficult to keep urban spaces adaptable to environmental constraints and economic, social and cultural changes. In other words, it has become difficult for conventional technologies to meet social demands for the upgrading of social capital in a sustainable manner and for the regeneration of attractive urban space that is not only safe and highly efficient but also conscious of historical, cultural and local identities to guarantee a high quality of life for all. Therefore, what is needed now is the creation of a new discipline that is able to reorganize the existing social capital and the technologies to implement it.

For this purpose, there is a need to go beyond the boundaries of conventional technologies of construction and structural design and to integrate the following technologies:

(1) Technology concerned with environmental and risk management
(2) Technology of conservation and regeneration with due consideration to the local characteristics of existing structures including historical and cultural resources
(3) Technologies of communication, consensus building, plan making and space management to coordinate and integrate the individual activities initiated by various actors of society

Up to now, architecture, civil engineering, and urban engineering in their respective fields have, while dealing with different time-space scales and structures, accumulated cutting-edge knowledge and contributed to the formation of favorable urban spaces. In the past, when emphasis was put on developing new residential areas and constructing new structures, development and advancement of such specialized disciplines were found to be the most effective.

However, current problems confronting urban development can be highlighted by the fact that a set of optimum solutions drawn from the best practices of each discipline is not necessarily the best solution. This is especially true where there are relationships of trade-offs among such issues as human risk and environmental load. In this way, the integration of the above three disciplines is strongly called for.

In order to create new integrated knowledge for sustainable urban regeneration, the Center for Sustainable Urban Regeneration (cSUR), The University of Tokyo, was established in 2003 as a core organization of one of the 21st Century Centers of Excellence Programs funded by the Ministry of Education and Science, Japan, and cSUR has coordinated international research alliances and collaboratively engages with common issues of sustainable urban regeneration.

The cSUR series are edited and published to present the achievements of our collaborative research and new integrated approaches toward sustainable urban regeneration.

Y. Fujino, T. Noguchi (Eds.)

Stock Management for Sustainable Urban Regeneration

 Springer

Yozo Fujino
Professor
Department of Civil Engineering
The University of Tokyo
7-3-1 Hongo, Bunkyo-ku, Tokyo
113-8656, Japan

Takafumi Noguchi
Associate Professor
Department of Architecture
The University of Tokyo
7-3-1, Hongo, Bunkyo-ku, Tokyo
113-8656, Japan

Cover photo: Roppongi Hills "Tokyo City View"; © Hironori Nagai

Library of Congress Control Number: 2008935036

ISSN 1865-8504
ISBN 978-4-431-74092-6   Springer Tokyo Berlin Heidelberg New York
e-ISBN 978-4-431-74093-3
DOI: 10.1007/978-4-431-74093-3

Printed on acid-free paper

Springer is a part of Springer Science+Business Media
springer.com
Printed in Japan

Preface

While the development of material civilization and industrialization initiated by the Industrial Revolution in the eighteenth century made possible a more comfortable life for mankind, it caused a concentration of human resources and led to rapid urbanization around the world. Consequently, a multitude of environmental problems such as global warming, disruption of the ecosystem, depletion of natural resources, and the accumulation of waste have become international issues. Urbanization accelerated the construction of super high-rise buildings, huge complex facilities, and stacked-up networks of roads and railways. At the same time, however, it resulted in crowded city blocks that were fragile and vulnerable to natural disasters such as earthquakes, tsunamis, and floods, and in historical structures becoming ruined and cultural urban space exhausted. Maintenance and repair of the amassed stock of structures are causing an economic burden today.

Against this background, the Urban Stock Management Research Group in the Center for Sustainable Urban Regeneration (cSUR) at the University of Tokyo has conducted research to develop integrated methods to maintain existing urban assets and to conserve the cultural/social context of urban environments. Urban stock management, such as conservation, utilization, and renovation, is vital for simultaneously supporting urban history, culture, changing lifestyles, and other conflicting elements. Therefore, new engineering for renovation and utilization of urban stock is urgently required to maintain safety, the environment, and continuing comfort as well as to save natural resources and reduce wastes.

This book is the crystallization of the knowledge and experience of the Urban Stock Management Research Group in the cSUR. We hope that the ideas and technologies included here will serve as useful and inspiring resources for researchers, practitioners, and students, and for those who are interested in the design, construction, maintenance, and recycling of urban structures to create sustainable cities and regions.

Yozo Fujino
Takafumi Noguchi

Acknowledgments

We would like to thank all the authors as well as those who participated in the research activities for their contributions, with special thanks to the following individuals for their editorial contributions.

Student assistants:
Tomoko Hirano, Ramin Motamed, Chikako Fujiyama,
Prince O'Neill Iqbal, Shohei Ueno, and Ke Wan Ling

Researchers:
Dionysius Manly Siringoringo and Ryoma Kitagaki

English editor:
Edward Moran

Contents

List of Contributors

Tomonari Yashiro
Professor
Institute of Industrial Science
The University of Tokyo
4-6-1 Komaba, Meguro-ku, Tokyo
153-8505, Japan
yashiro@iis.u-tokyo.ac.jp

Mikio Koshihara
Associate Professor
Institute of Industrial Science
The University of Tokyo
4-6-1 Komaba, Meguro-ku, Tokyo
153-8505, Japan
kos@iis.u-tokyo.ac.jp

Keisuke Fujii
Associate Professor
Department of Architecture
The University of Tokyo
7-3-1 Hongo, Bunkyo-ku, Tokyo
113-8656, Japan
tkfujii@mail.ecc.u-tokyo.ac.jp

   Arranged and Translated
   by Mizuko Ugo

Shin Muramatsu
Professor
Institute of Industrial Science
The University of Tokyo
4-6-1 Komaba, Meguro-ku, Tokyo
153-8505, Japan
muramatsushin@aol.com

Tetsuya Ishida
Associate Professor
Department of Civil Engineering
The University of Tokyo
7-3-1 Hongo, Bunkyo-ku, Tokyo
113-8656, Japan
tetsuya.ishida@civil.t.u-tokyo.ac.jp

Takafumi Noguchi
Associate Professor
Department of Architecture
The University of Tokyo
7-3-1, Hongo, Bunkyo-ku, Tokyo
113-8656, Japan
noguchi@bme.arch.t.u-tokyo.ac.jp

Muneo Hori
Professor
Earthquake Research Institute
The University of Tokyo
1-1-1 Yayoi, Bunkyo-ku, Tokyo
113-0032, Japan
hori@eri.u-tokyo.ac.jp

Ken'ichi Kawaguchi
Professor
Institute of Industrial Science
The University of Tokyo
4-6-1 Komaba, Meguro-ku, Tokyo
153-8505, Japan
kawaken@iis.u-tokyo.ac.jp

Hitoshi Kuwamura
Professor
Department of Architecture
The University of Tokyo
7-3-1 Hongo, Bunkyo-ku, Tokyo
113-8656, Japan
kuwamura@arch.t.u-tokyo.ac.jp

Koichi Maekawa
Professor
Department of Civil Engineering
The University of Tokyo
7-3-1 Hongo, Bunkyo-ku, Tokyo
113-8656, Japan
maekawa@concrete.t.u-tokyo.ac.jp

Yukio Nishimura
Professor
Department of Urban Engineering
The University of Tokyo
7-3-1 Hongo, Bunkyo-ku, Tokyo
113-8656, Japan
nishimur@ud.t.u-tokyo.ac.jp

Kazumasa Ozawa
Professor
Department of Civil Engineering
The University of Tokyo
7-3-1 Hongo, Bunkyo-ku, Tokyo
113-8656, Japan
ozawa@ken-mgt.t.u-tokyo.ac.jp

Yozo Fujino
Professor
Department of Civil Engineering
The University of Tokyo
7-3-1 Hongo, Bunkyo-ku, Tokyo
113-8656, Japan
fujino@civil.t.u-tokyo.ac.jp

Tsuyoshi Takada
Professor
Department of Architecture
The University of Tokyo
7-3-1 Hongo, Bunkyo-ku, Tokyo
113-8656, Japan
takada@load.arch.t.u-tokyo.ac.jp

Ikuo Towhata
Professor
Department of Civil Engineering
The University of Tokyo
7-3-1 Hongo Bunkyo-ku, Tokyo
113-8656, Japan
towhata@geot.t.u-tokyo.ac.jp

1. Stock Management of Civil Engineering Infrastructure in Asia

Kazumasa Ozawa

1.1 Infrastructure Development in Asia

1.1.1 Introduction

Infrastructure development, which can help distribute resources and serve as a base of service delivery, is indispensable to the welfare of human society and development. The quality and efficiency of infrastructure have a great influence on the quality of life of the people, social soundness, and economic activity. The Romans, for example, developed an extensive road network, built aqueducts and bridges, and thus helped cement their empire. The provision of infrastructure supported the economic development of every country in the world, and superior infrastructure systems have contributed greatly to an improvement in the overall quality of life. This paper is designed to introduce some views on infrastructure management and stock management in the countries of Asia. It focuses on management aspects of the civil engineering infrastructure, i.e. it excludes buildings.

Figure 1-1 shows the economic development in Asian countries (the gross national income per capita from 1960 to the present). Compared to Southeast Asia and South Central Asia, East Asia has experienced a far faster economic growth over this period. Infrastructure is indispensable to the welfare of human society as well as economic development.

The transition of social capital stock in Japan, based on the data from the Cabinet Office is described in Fig. 1-2. In 1998, it was valued at more than 600 trillion yen. The major parts of it are in the categories of "Road" and "Mountain/River Improvement." Economic development in Japan has also been supported by integrated infrastructure development.

Y. Fujino, T. Noguchi (eds.) *Stock Management for Sustainable Urban Regeneration*,
© 2009 to the complete printed work by Springer, except as noted. Individual authors
or their assignees retain rights to their respective contributions; reproduced by permission.

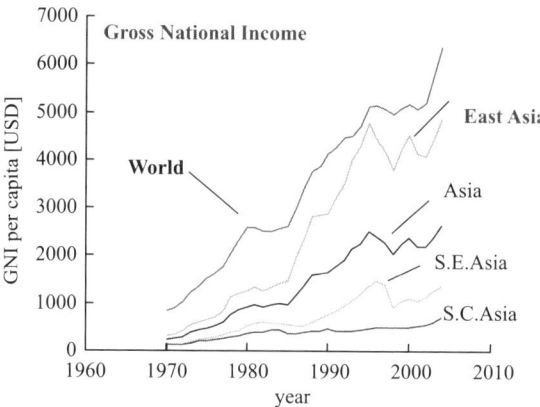

Fig. 1-1. Economic development in East Asia (Fujino (2006))

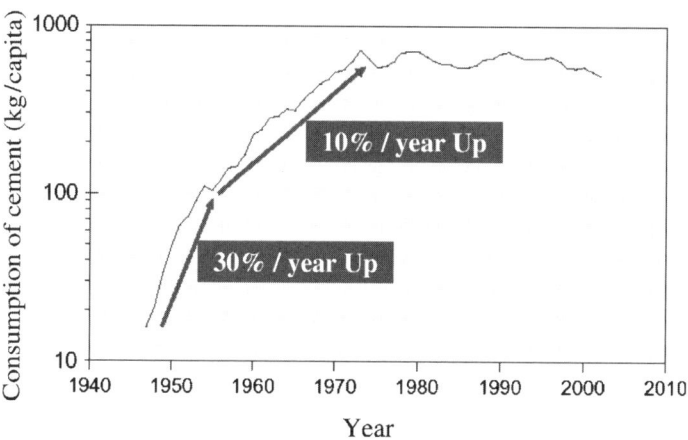

Fig. 1-2. Consumption of cement per capita in Japan: 1947–2003 (logarithmic scale) (Ouchi (2004))

1.1.2 Infrastructure Development and Economic Growth

From the point of view of the amount of infrastructure, such as concrete structures, constructed in each country, valuation can be based on the consumption of cement, because the volume of cement is usually proportional to the volume of the concrete.

Figure 1-3 shows the consumption of cement per capita in Japan (1947–2003) in a logarithmic scale. Immediately after WWII, there had been a rapid growth (around 30 percent annually) in consumption for about 10 years. During the next 20 years, there still was almost 10 percent growth

Brunei	2,515	Jordan	560	Bhutan	149
Qatar	2,124	Oman	541	Vietnam	147
UAE	2,072	Malaysia	506	Sri Lanka	124
Kuwait	1,050	Turkey	468	Indonesia	106
South Korea	1,015	China	463	Iraq	105
Taiwan	818	Macao	457	India	99
Saudi Arabia	806	Iran	346	Pakistan	73
Singapore	785	Syria	319	Cambodia	57
Israel	748	Thailand	285	Mongolian	52
Lebanon	695	North Korea	202	Myanmar	45
Hong Kong	677	Laos	191	Nepal	44
Bahrain	651	Yemen	170	Bangladesh	39
Japan	570	Philippines	157	Afghanistan	37

Fig. 1-3. Cement consumption per capita in Asian countries (Ouchi (2004))

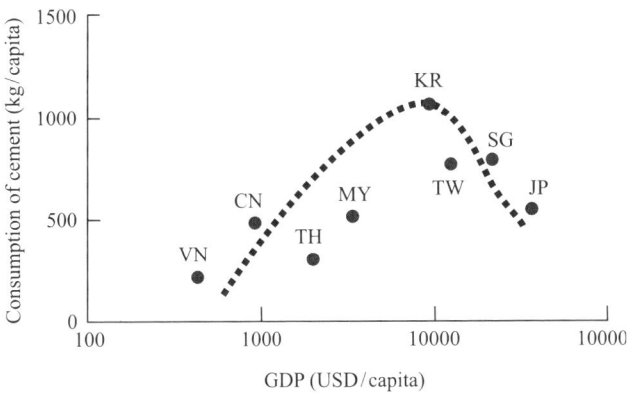

Fig. 1-4. Relationship between GDP and Consumption of cement in Asian countries (Ouchi (2004))

in terms of annual cement consumption. In the 1970s, the peak came and the trend became almost flat. Nowadays it is gradually decreasing. This shows a pattern of the scenario of the stock of infrastructure, a trend that mirrors that of economic development.

The consumption of cement in Asian countries in 2000 is described in Fig. 1-4. The top four countries are Brunei, Qatar, UAE and Kuwait, all West Asian countries. Then comes South Korea and Taiwan. Japan is in the 13th place of all 39 countries. Around the middle of the list are found Malaysia, Turkey, China, and Thailand. At the bottom are Mongolia, Myanmar, Nepal, and Bangladesh. Finally, Afghanistan is at the last of the list.

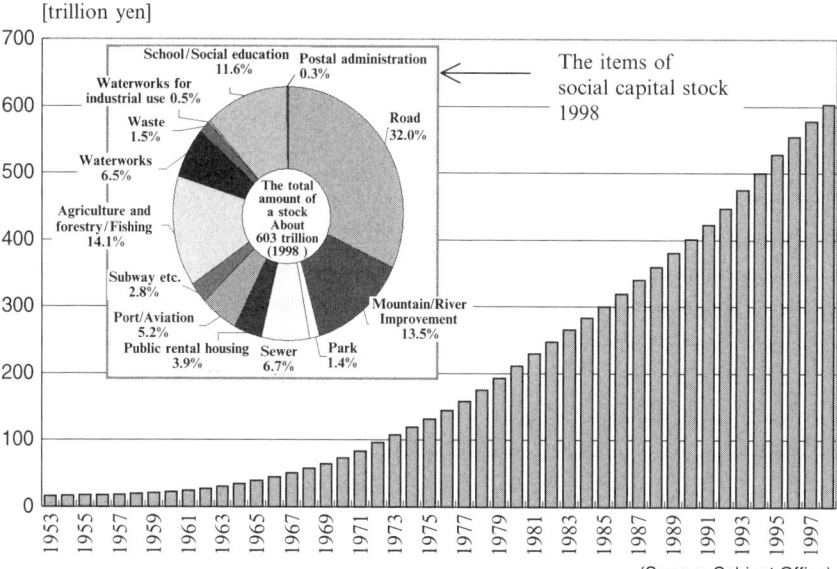

Fig. 1-5. Transition of social capital stock in Japan

(Source: Cabinet Office)

These countries at the bottom will experience a larger cement consumption in the future, along with their economic growth. Then the consumption will gradually decrease in other countries. This is the general scenario of cement consumption in terms of infrastructure development.

Figure 1-5 illustrates the relationship between GDP per capita and the cement consumption per capita in 2000. Basically the cement consumption increases with the increase of GDP up to the peak of 10,000 to 20,000 USD per capita and decreases with a GDP of 20,000 USD per capita. Vietnam, China, Thailand, and Myanmar are regarded as countries whose infrastructure development is on the rise; on the other hand, Taiwan, Singapore, and Japan are on a descending curve of cement consumption even if they can show economic growth. Korea is at the peak of cement consumption and may follow the trend of Singapore and Japan.

1.1.3 Considerations for Maintenance

As for the age of bridges in Japan, the peak is around 30 to 40 years old as shown in Fig. 1-6. So 20 years from now, these bridges will be 50 to 60 years old. Soon we will face the rapid increase of the number of bridges more than 50 years old; now we have 2000; in 5 years, this number will be doubled; in 10 years, five times greater; and in 15 years, by ten times greater.

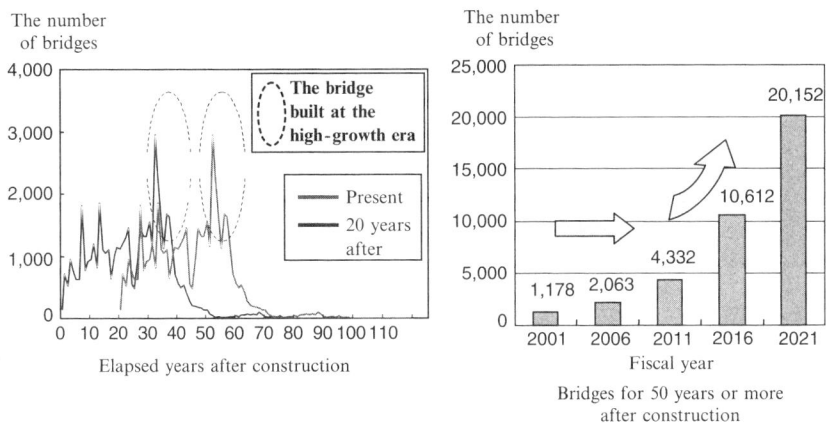

Fig. 1-6. Situation of bridges (National highway and four public corporations)

This is because we made a concentration of investment in the bridges about 30 to 40 years ago. In many Asian countries, especially the mega cities, a similar concentration of investment has been made to their infrastructure. Some years later, their infrastructure will enter into old age at the same time. So we have to consider how to tackle the problems of maintenance and rehabilitation for much of this old infrastructure. This is something that should be considered during the construction of new structures.

In Fig. 1-7 are shown damaged bridges that are owned by the Bangkok Metropolitan Authority, Bangkok, Thailand. They need to invest in new infrastructure, but they also have to rehabilitate the damaged structures.

Concerning the case of the water supply system in Manila, Philippines, as shown in Fig. 1-8, there is a need to invest in new construction in order to enlarge the water supply in metropolitan Manila, but they also need to take care of the maintenance and rehabilitation of the existing system, which was established in 1878.

Non-revenue water (the amount of water which can be regarded as non-revenue divided by the total amount of supplied water) accounts for about 50 percent of the total. Half of the water disappears, largely because of the leakage from old pipes, which is a very critical issue for the water supply company.

1.1.4 Problems of Infrastructure Management

The transition of the maintenance cost of bridges in a certain prefecture in Japan is shown in Fig. 1-9. Due to financial difficulties, a local government

Fig. 1-7. Bridges in Bangkok, Thailand

has recently reduced new construction costs as well as maintenance costs even though the total infrastructure has increased. Financial difficulties in local governments such as prefectures, cities, towns and villages are in more severe condition than that of the central government.

On the other hand, the situations of maintenance support systems in local government are poor. Based on the data from the questionnaire survey by a JSCE committee in 2003, few cities, large or small, inspected their infrastructure as shown in Fig. 1-10. And many local governments did not make specific maintenance plans, which are usually based on the inspection data

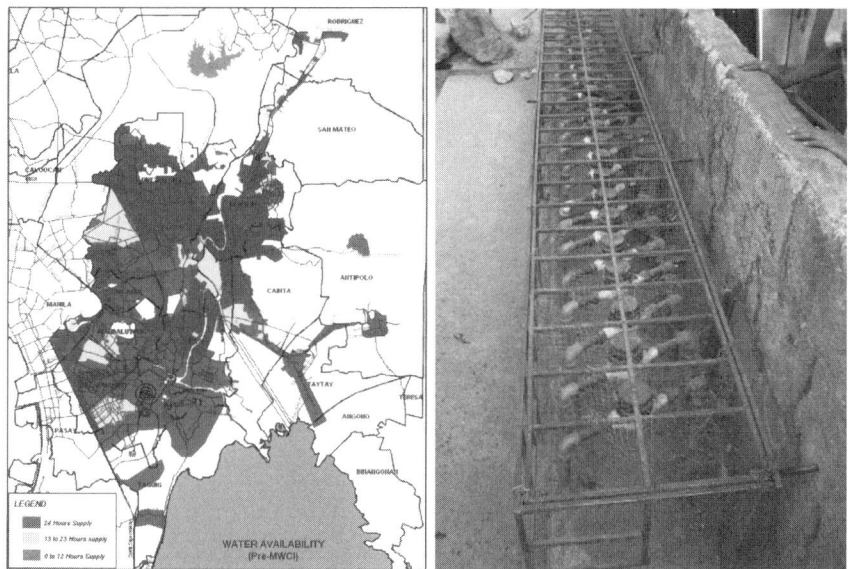

Non Revenue Water 50%
Water system was originally established in1878. (MWCI)

Fig. 1-8. Water supply in Manilla, Philippines

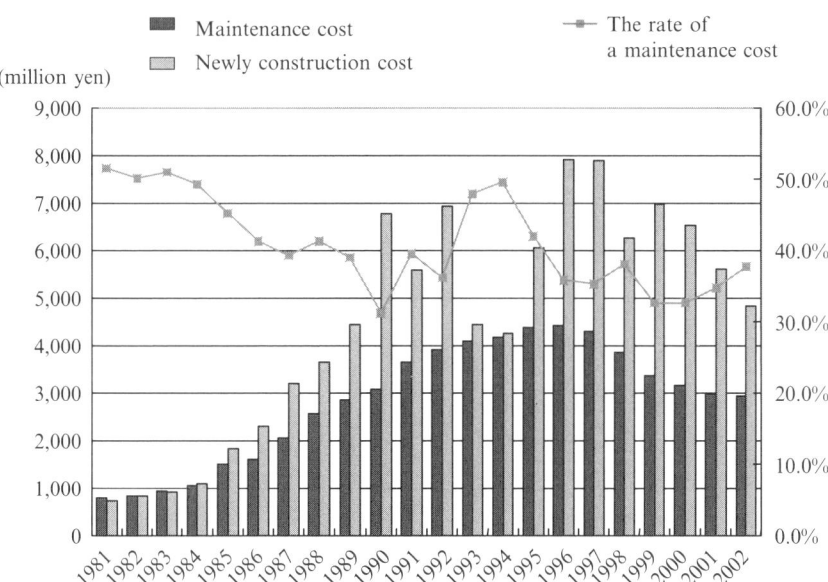

Fig. 1-9. Transition of the maintenance cost of the bridge in a certain prefecture in Japan

	Inspection			Maintenance plan		
	Do	A part is carried out	No	Do	Under consideration	No
Prefectures	12 (33%)	6 (17%)	18 (50%)	9 (27%)	21 (64%)	3 (9%)
Ordinance designated city	6 (86%)	0 (0%)	1 (14%)	0 (0%)	6 (86%)	1 (14%)
Large City (300,000 people)	10 (43%)	0 (0%)	13 (57%)	1 (4%)	8 (35%)	14 (61%)
Small City (50,000 people)	7 (33%)	0 (0%)	14 (67%)	0 (0%)	6 (30%)	14 (70%)

(source: questionnaire survey by JSCE committee)

Fig. 1-10. Situation on maintenance in Japan (August 2003)

and the actual situation of the infrastructure. Almost all of the local governments have some statistical information on infrastructure and investment, but many of them did not have any repair history or inspection-results data according to the questionnaire survey.

So the problem of infrastructure management is due to an increase of aging facilities, the low priority of maintenance budgets, the lack of long-term plans for maintenance, the lack of data, and the problems of the organization for maintenance.

East Asian countries may be facing such problems more seriously due to the concentration of investment for infrastructure development as a means to rapid economic growth.

1.2 Asset Management for Infrastructure

1.2.1 Concept of Asset Management

In order to solve the problems of infrastructure management, the concept of asset management for infrastructure is proposed. This is defined as a set of systematized activities for managing infrastructure efficiency and effectively in the long term, using the knowledge in fields such as engineering, economics, and business administration.

The expected trajectory of such asset management is fourfold: first, rational budgetary request to the financial sector based on future projections;

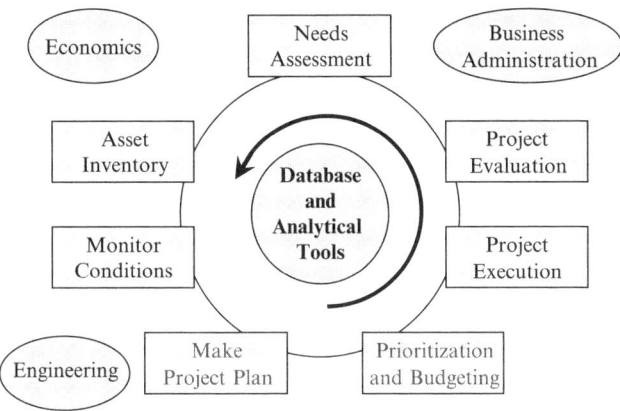

Fig. 1-11. Cycle of infrastructure management

second, the improvement of asset conditions; third, effective use of the public funds according to the life cycle cost concept; and finally improvement in accountability.

The cycle of infrastructure management based on the concept of asset management is shown in Fig. 1-11. Need assessment, asset inventory, monitoring of conditions, developing a project plan, prioritization and budgeting, project execution and project evaluation are all necessary in the process of infrastructure management. This process should also be carried out based on knowledge in the fields of engineering as well as economics and business administration.

For instance, in order to make a project plan of maintenance some key issues should be included. Of first importance is the selection of a scenario of maintenance that considers total life cycle costs. To compute these costs, a future projection of the deterioration of infrastructure is needed, one that considers the repair effect. The second step is prioritization of a project plan. Because one is usually faced with budget restrictions, every project plan cannot be achieved, so they have to be selected and prioritized taking into consideration the annual equalization of the budget, based on the equalization concept of sharing the burden between generations. Finally, the adjustment to individual project plans is important after the prioritization and selection of the plans. Original project plans should be changed in accordance with degree of damage, risk and asset value because not every project plan can be done at the same time. Knowledge in the fields of engineering as well as economics and business administration is needed as a prerequisite to this process.

1.2.2 System for Infrastructure Management

In order to realize the idea of asset management, it is necessary to build a system for infrastructure management. The whole system for infrastructure management is shown in Fig. 1-12. In the center, the management system of physical property is located according to the cycle of infrastructure management.

It is also necessary to prepare the management system for the funds required for project implementation, the management of an organization and a staff, and the analytical skills and the database that support various judgments. As for the management of funds, it should be considered that the range of budget allotment and the utilization of private financing techniques might be necessary when public sources of revenue run short. In terms of organization and staff management, it is important to establish a policy that includes provision both for an internal staff and outsourcing, and that describes how to train the in-house engineers. In implementing technical development and a database, it is necessary to decide what kind of technical development is required, and what kind of procedure is required for a useful database. Neither can be realized immediately and it is important to make the strategy to perform one step at a time with a consideration of sustainability.

Public infrastructure is a public asset. The central and local government has the responsibility of explaining the investment and the expense for maintenance of infrastructure to the public. An accounting system for infrastructure management, based on physical evaluation, book-value evaluation

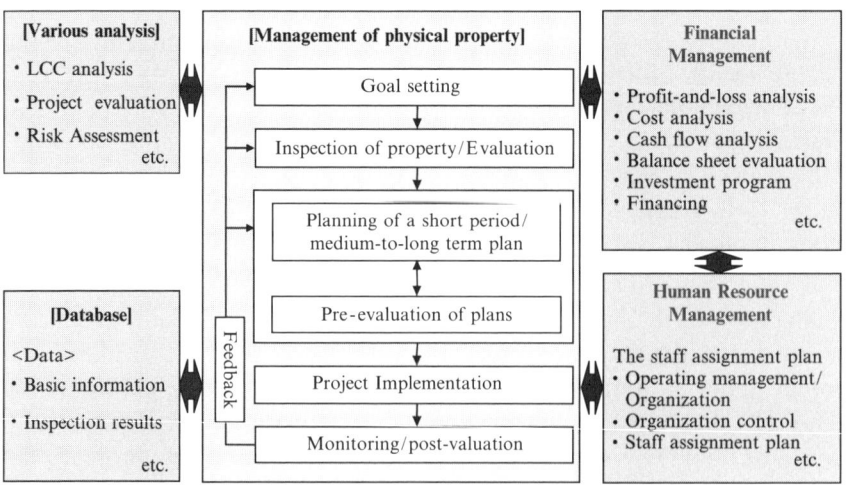

Fig. 1-12. System for infrastructure management

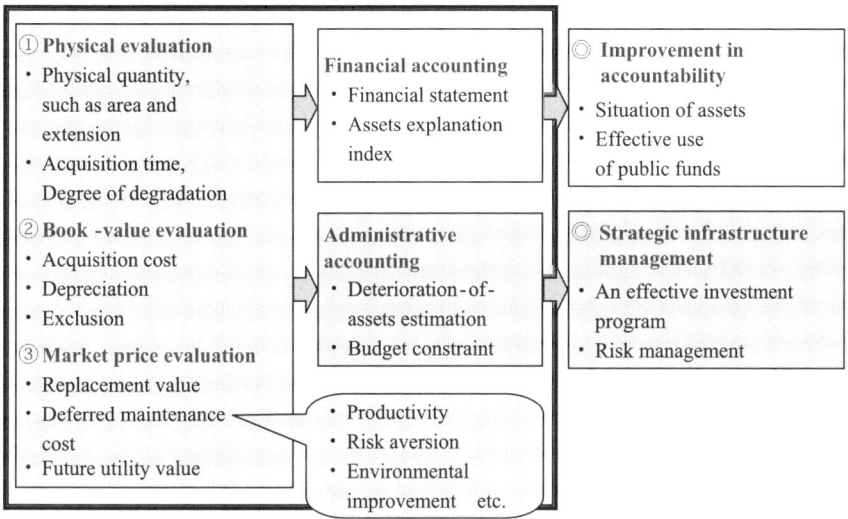

Fig. 1-13. Accounting system for infrastructure management

and market price evaluation is useful for financial accounting and adminis-trative accounting as shown in Fig. 1-13. Using this kind of system, public accountability of infrastructure management can be improved and some strategic infrastructure management can be achieved.

The example of a balance sheet for infrastructure is shown in Fig. 1-14, where the asset values of pavement, bridges, tunnels, banks and land are described. Also given in the balance sheet are the amount of increase of assets in some years; acquisition and updating, exclusion, depreciation, and the average life and average property age of stock. Asset values and asset conditions are thus available to the public every year using such a balance sheet.

For the purpose of strategic management, a simulation of the transition of the amount of valuation of asset is based on a future projection of asset values. Thus the transition of the physical condition of the whole property can be also described.

1.2.3 Financing for Infrastructure Management

It is necessary to arrange financial management for better infrastructure management. Road maintenance and road construction or rehabilitation is usually covered by the general budget. But this is sometimes not enough for sufficient maintenance of roads.

(※Data is the imaginary value)   Unit: 10,000 yen

| Kind | Asset value | The amount of increase | | | | Depreciation total | Average life | Average property age |
		from 2002	Acquisition · updating	Exclusion	Depreciation			
Pavement	754,820	7,490	57,970	▲6,080	▲44,400	1,012,860	30	17
Bridge	242,070	3,230	6,720	▲390	▲3,100	894,210	100	78
Tunnel	188,700	▲5,100	300	0	▲5,400	440,300	50	35
Bank	532,330	1,320	9,180	▲30	▲7,830	1,131,200	100	68
· · ·	· · ·	· · ·	· · ·	· · ·	· · ·	· · ·	· · ·	· · ·
Land	1,735,200	21,200	21,200	–	–	–	–	–
· · ·	· · ·	· · ·	· · ·	· · ·	· · ·	· · ·	· · ·	· · ·
Total	3,618,400	31,230	101,370	▲9,110	▲61,030	3,871,350	–	–

Average property age =
    Depreciation total/(Asset value + Depreciation total) × Average life

Fig. 1-14. The example of evaluation of infrastructure asset

		Maintenance Expenditure/GDP% (1970–2003)	Maintenance Expenditure/GDP% (1990–2003)
Developing Countries with Road Fund	Hungary	1.35	0.17
	Latvia	1.44	1.44
	Romania	0.20	0.20
	Ethiopia	Average=0.61 0.24	0.58 0.14
	Benin	0.25	0.16
	Yemen	1.62	1.86
	Ecuador	0.22	0.22
	Ukraine	0.44	0.44
Developing Countries without Road Fund	Sri Lanka	0.0569	0.0031
	Pakistan	0.2600	0.2254
	Albania	Average=0.20 0.0542	0.18 0.0542
	Croatia	0.1741	0.1626
	Bolivia	0.5172	0.4833
	Mongolia	0.1344	0.1344
Middle Income Countries without Road Fund	Thailand	0.21	0.210
	Philippines	0.29	0.140
	Poland	0.52	0.510
	Turkey	Average=0.26 0.13	0.26 0.130
	Chile	0.20	0.420
	Brazil	0.10	0.014
	Hong Kong	0.49	0.520
	Mexico	0.15	0.130

Source: Road expenditure data are from World Road Statistics, IRF and GDP data from World Bank Statistics
(S. Nirosha Malkanthi)

Fig. 1-15. Developing and middle income countries: maintenance expenditure on roads as proportion of GDP

Some countries have introduced a road fund for such maintenance collected from users via toll charges and tax on fuels.

It is obvious that maintenance expenditures in a country with road funds are greater than that without such funding, as seen in Fig. 1-15. For developing countries with road funds, the average maintenance expenditure per GDP is more than 0.5 percent. But for developing and middle-income countries without road funds, the average is only about 0.2 percent. The countries with road funds make more investment toward the maintenance of roads. It is important to establish a fund for maintenance because it is easier to cut the expenditure for maintenance in the case of financial difficulties.

1.3 Concluding Remarks

Infrastructure development has an important role on economic growth in Asia. The recent rapid economic growth especially in East Asia created a concentration of investment in infrastructure development. It can be estimated that rehabilitation and updating of deteriorated structures will be needed at the same time in the future.

It is necessary to establish a better infrastructure management system, termed asset management, to improve the asset conditions, to use public funds effectively and to improve accountability.

In order to construct and to carry out an infrastructure management system, it is necessary to establish a management cycle for physical assets as well as to prepare financial management, human resource management, database and analytical tools. An accounting system for infrastructure management is also useful to improve accountability and strategic management for all of the infrastructure assets.

It is indispensable to construct an infrastructure management system to extend the life of the infrastructure, reduce maintenance and rehabilitation costs and better serve the public service, which will lead sustainable development in Asia in the future.

References

Fujino, Y. (2006) Monitoring of bridges and transportation infrastructure. Proceedings of EASEC-10: 8

Kazumasa, O. (2004) *Infrastructure Management and Life-Cycle Cost-Assessment*, in Proceedings of the First International Conference on Construction Information Technology, Beijing, China, pp.148–154

Malkanthi, S.N. (2006) *Study of Effectiveness of Road Fund as a Solution for Maintenance Problems in Sri Lanka* (Master's thesis for University of Tokyo)

Ouchi, M. (2004) "Influence of economic growth on the consumption of cement", *Journal of Construction Management (JSCE)* 11: 249–260

Sub-committee on Asset Management for Infrastructure (2005) Challenge to the asset management. Japan Society of Civil Engineers

2. Overview of Building Stock Management in Japan

Tomonari Yashiro

2.1 Introduction

For the last six decades, construction activities in Japan have mainly focused on new building activities rather than work on existing buildings and facilities. For several reasons, there has been a serious shortage of buildings and facilities with sufficient quality and performance: destruction during World War II coupled with the huge demand for new construction by the concentration of population in mega cities since the 1950s.

However, Japan is now faced with the situation of managing existing building stocks and facilities which have been accumulating for the last six decades. It requires a fundamental paradigm shift from new-building-based thinking to a model of thinking about working on existing buildings. Asian countries that are experiencing enormous new building activities will inevitably face the similar situation that Japan now faces. Therefore, the Japanese situation could provide lessons for future construction activities in Asia.

As an introduction to this book, this chapter describes why Japan needs to establish methods of stock management from the aspect of macro-scale building stock formation. The chapter also presents the potential of information provisions using information technology (IT) as an enabler for a holistic approach in stock management.

Y. Fujino, T. Noguchi (eds.) *Stock Management for Sustainable Urban Regeneration*,
© 2009 to the complete printed work by Springer, except as noted. Individual authors
or their assignees retain rights to their respective contributions; reproduced by permission.

2.2 Why Stock Management?

2.2.1 Age Distribution of Building Stock

Fig. 2-1 shows how much building stock has been accumulated in the last six decades in Japan. Japan has now over 8 billion square meters of building stock. Fig. 2-1 seems to show a very healthy, gradual increase in the national building stock.

Fig. 2-2 shows that the Japanese building stock is quite relatively new, with most of the building stock has been constructed after the 1970s; as a matter of fact, 80 percent of the buildings now in use have been constructed after that decade. The buildings constructed before that time represent less than 20 percent of the whole stock. Some work that has been done to these buildings, such as retrofitting, refurbishment, redesign or remodeling, has happened 20–30 years after their construction. Thus, it is presumed that a huge potential number of buildings will be targeted for building refurbishment in the near future.

Fig. 2-3 shows a comparative age distribution of existing buildings in the European and the North American countries. In Germany, France and England the postwar buildings occupy 50–60 percent of the building stock, while in Japan most of the building stock has been constructed after the Second World War. It is clear that the Japanese building stock is relatively new while it is

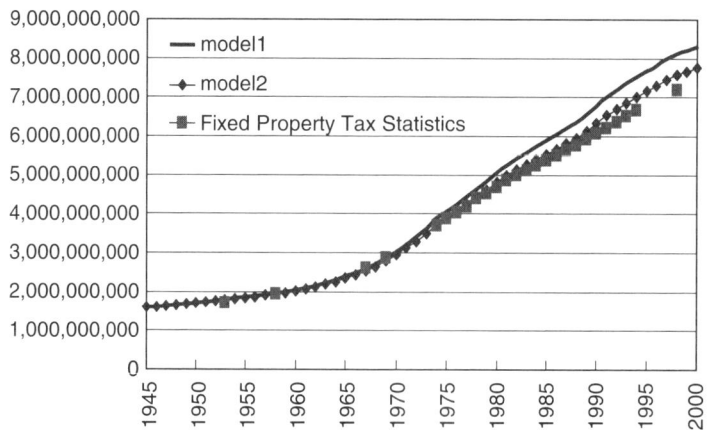

Fig. 2-1. Increase of building stock in Japan

Age distribution of building stocks in Japan

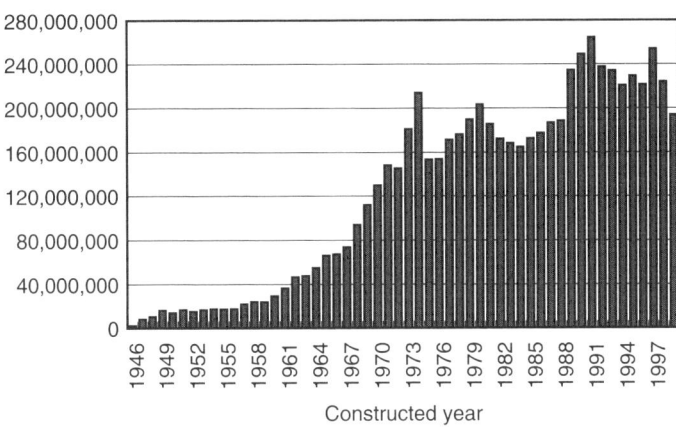

Total Floor Area (m sq.)

Fig. 2-2. Age distribution of housing stock in Japan

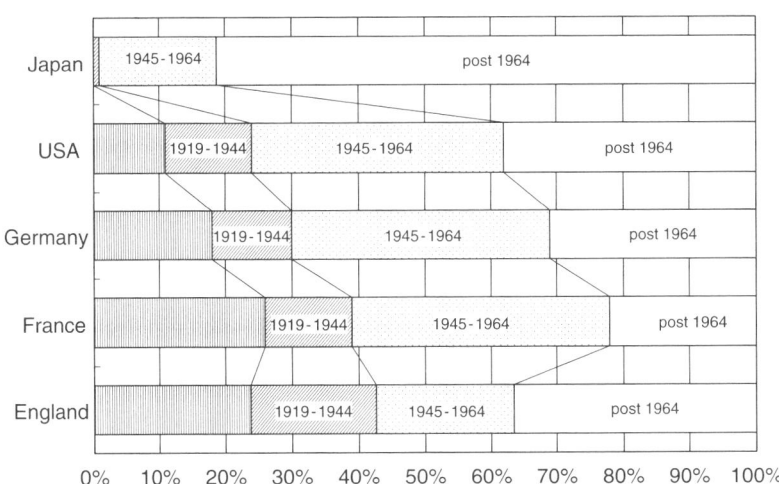

Age composition of building stocks in several countries
(Japan: Investigated by Yashiro, T. + Others by English House Condition Survey)
Japan's composition is in a process of rapid aging.

Fig. 2-3. Comparison of age distribution of housing stock (Source: English House Condition Survey, for Japan as calculated by the author)

older especially in the European countries. In England for instance, almost 40 percent of the building stock constructed before 1945 is still in use.

2.2.2 Short Building-Life Syndrome

Fig. 2-4 describes that in several cases, buildings in Japan are very new and thus exhibit the problem of the "short buildings-life syndrome." In the charts, the X axis shows the years since construction and the Y axis shows the percentage of buildings that have survived. The number of steel structural office buildings is decreasing because of demolition; within less than 30 years over 50 percent of these buildings will have been demolished.

Thus, the average life span for steel-structured buildings is less than about 30 years while it is almost 40 years for reinforced concrete buildings. This is a rather short span because statistics show that the life span of small timber houses in Japan is about 40 years (Fig. 2-5). After some surveys by colleagues, the middle sample indicates that the life span for small timber houses is now about 50 years.

In the United States, taking Indianapolis as an example (Fig. 2-5), statistics show that the average life span of a house is over 100 years. By comparison,

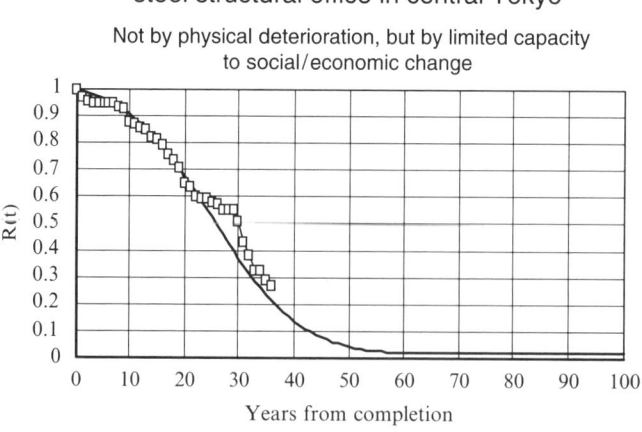

Fig. 2-4. Short-building-life syndrome in Japan for steel structure buildings and reinforced concrete buildings

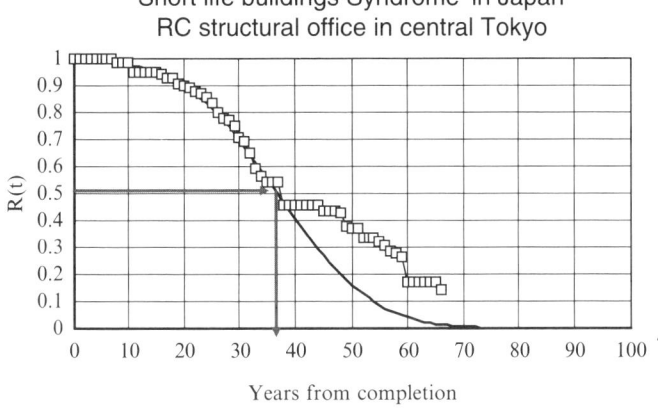

Fig. 2-5. Statistical analysis on house life in Japan and USA

Japanese houses have a much shorter life span. This is not due to physical deterioration though about 11 million houses do not meet with the current seismic building code, so it is probable that millions of houses will be destroyed by earthquakes. However, even if a building itself meets structural safety codes, it can still be in use only for 50 years simply because it does not meet with ever-changing social and economic requirements. The life of many Japanese buildings is short mainly due to socioeconomic issues rather than to reasons of physical integrity. So when one talks about the longer life of buildings, a system needs to be developed that integrates technological as well as socioeconomic issues.

2.2.3 Decline of New Building Demand

Fig. 2-6 shows the ratio between newly built construction and the real increase of the building stock. Some buildings are demolished and replaced by new ones. In that case the building stock both decreases and increases, therefore any new construction is not equal to the new increase of the building stock itself. Until the end of the 1990s, almost 80 percent of new construction contributed to the increase of the building stock, but after the 1990s that ratio went down to less than 40 percent. This means that in Japan some 150 million square meters are constructed per year but only 60–70 million of these square meters result in an increase of the building stock. This fact suggests that the demand for new building space is declining in Japan.

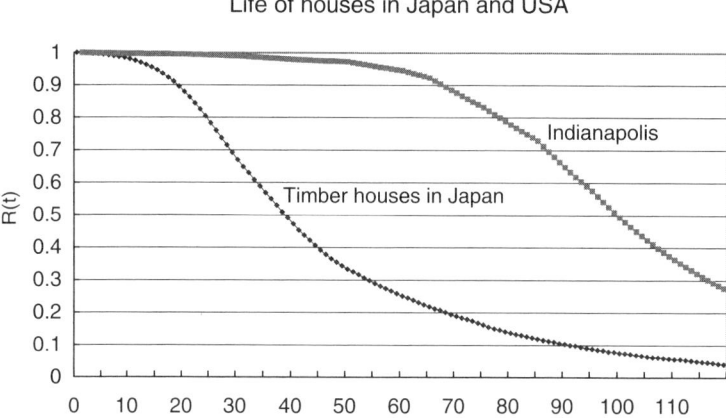

Fig. 2-6. Decline in demand of floor space

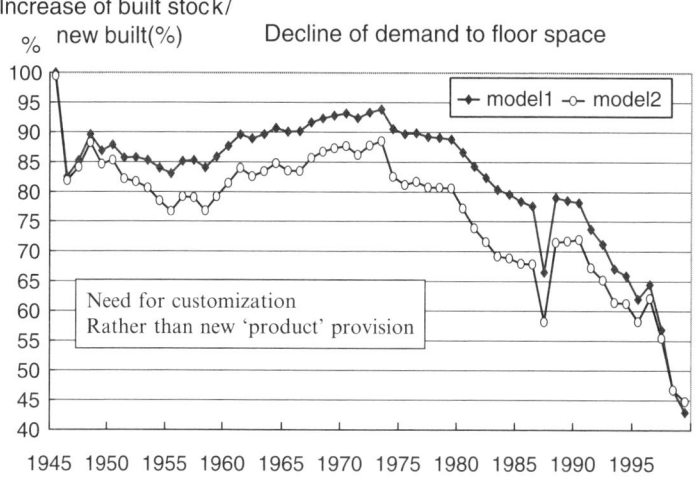

Fig. 2-7. Number of households in England, by year of house construction (Source: English House Condition Survey)

2.2.4 Comparative Content of Housing Stock in UK and Japan

A very comparative situation exists in the United Kingdom as shown in Fig. 2-7, which presents national housing statistics indicating the number of households living in structures built in given years. For example, the lowest graph

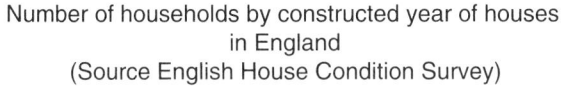

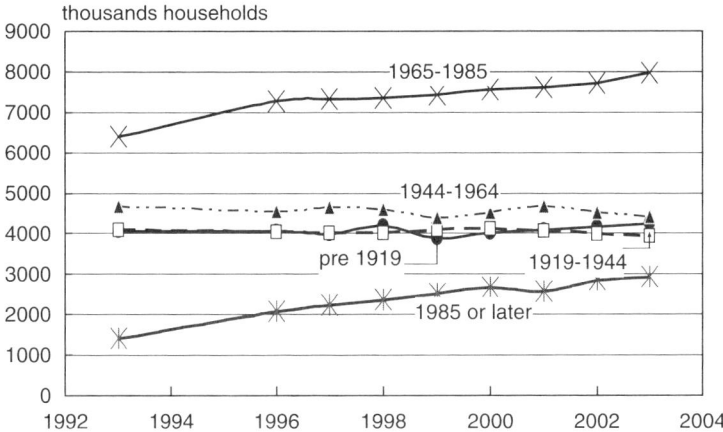

Fig. 2-8. Potential risk of households in Japan

shows households living in houses constructed after 1985 and the middle one shows the number of households living in houses constructed before 1919. This graph is almost horizontal, which means that, in England, the houses in every age cohort have not vanished so much. I heard from a British scholar that the life of a building in England is about 1,000 years because only 2,000 houses are demolished while about 160–200 thousand are newly constructed to accommodate the increase in households (Meikle and Connaughton (1994)).

Comparatively, the situation in Japan is explained in Fig. 2-8, in which the graph shows the amount of debt remaining in terms of the number of months. The "price" of houses (i.e. their value) starts to elapse right after the loan is obtained and construction begins. Anyone who has incurred a housing loan pays back this loan on a monthly basis, and due to the interest rates the graph has an upper hill shape. The significant thing in Japan is that the economic value in the market of timber houses is dropping down, and within 15 years the value will drop down between 10 to 20 percent of the initial cost.

In other words, if someone does not have any income then there is no way he or she can pay the house loan, so naturally the house will be sold, but at a lower value than when it was initially bought. Still, even if the house is sold there is still the remainder of the housing loan to be dealt with. This is one invisible risk. Earlier in time, very few Japanese would have recognized this invisible risk because there was a capital gain from land.

The comparative situation in England and in Japan in terms of the design life of the building versus its physical life is as follows: in Japan the design life is longer than the actual building's life. In the UK, and probably in other European countries, the design life is shorter than actual life; therefore, Europeans make some investments to rehabilitate the structure and so on. Japanese homeowners suffer from the invisible risk of losing the value of the housing investment while in England the house value can be maintained if the residents make investments in that house. For the last 10 years England enjoyed a good economic situation; therefore, the housing prices increased nearly fourfold during the last 20 years. So the people who bought houses 20 years ago are now enjoying this fourfold increase while the Japanese who invested in buying a house have almost lost their wealth. These are quite serious differences.

2.2.5 "Poverty Trap"

Fig. 2-9 illustrates a situation in which Japan now finds itself. It can be called the "poverty trap." When buying a newly built house, one typically takes on a 20- to 30-year mortgage; in Japan, once this loan is paid off, a male purchaser will enjoy 10 or 15 years left of his life (knowing that the average life of a Japanese man is about 70–80 years). His house would then be inherited by one of his children, Since the inheritance taxes are quite high, it is very probable that this son or daughter would sell the house, which eventually would be demolished. This leads the son or daughter absurdly back to square one, from where his/her parent had started. It

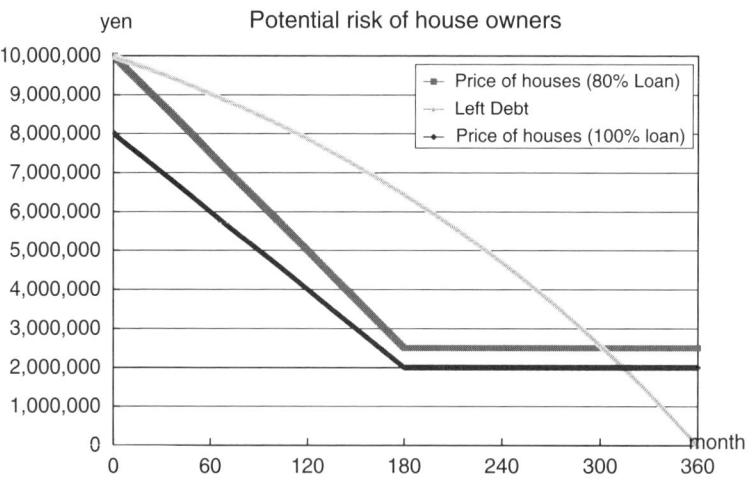

Fig. 2-9. Need of escape from "poverty trap" caused by rapid drop of house value by age associated with short-building-life syndrome

should be emphasized that there is a need to escape from this "poverty trap" situation by a restructuring of the technological and socioeconomic system that could enhance the value of existing buildings through continual investment. Stock management should take a major role in this restructuring.

2.2.6 Expected Paradigm Shift

Fig. 2-10 introduces the concept of life cycle value. It is the integration of the value over time on the buildings. The vertically hatched area of the graph represents the life cycle value. When new building activities are dominant in construction activities, most building engineers in Japan tend to only think about initial price or cost; consequently the life cycle value drops down through time. The building engineers and all the stakeholders who are involved in the building process need to completely change their way of thinking and jointly create a new paradigm that will maximize the value of a building over time. If the value of houses can be maintained in Japan or in Asia, the people here can enjoy the lifestyle that will allow them, once they buy a house in young age, to obtain economic stability throughout their entire lives.

Fig. 2-11 presents a comparison of resource productivity in a new-building-oriented society to an existing-buildings-oriented society. In a new-building-oriented society various resources are consumed and huge amounts of waste are generated while only very little value is created.

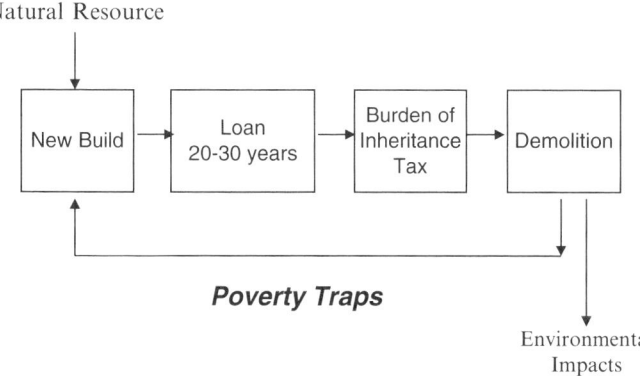

Fig. 2-10. Expected paradigm shift from initial value to life cycle value

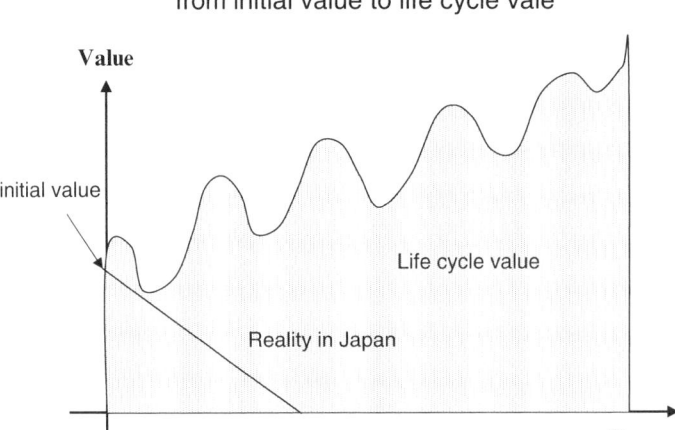

Fig. 2-11. New-building- oriented society vs Stock-based society

Contrarily, works to existing buildings involving repetitive use of the stock there will result in a consumption of fewer resources and less waste while adding more value to the existing stock.

2.2.7 Adaptability to Ever-Changing Requirements

Fig. 2-12 introduces the idea of open building, where the building is separated into two parts: the skeleton and the infill. The skeleton is the infrastructure of building and includes the structure, stairs and vertical pipes and wires that are commonly shared by all users of building. The infill is the private part of building that includes interior finishing and appliances. Thus, infill is expected to be exchangeable, corresponding to the ever-changing requirements of occupants and users while the skeleton lasts for 100 or 200 years with a well-planned long-term maintenance program.

The cycle of change is completely different for each of these two elements. Sadly, in most buildings, the two are entangled: both the element that is expected to last for 100 years and the element that needs to be replaced within 5–10 years are mixed together. It is difficult to sort them out. Consequently when the building is refurbished, the healthy part also inevitably needs to be destroyed.

The concept of "open building" is the principle and method of construction to assure full autonomy to both the skeleton and infill so that decisions can be made about each of them separately, as part of an independent package.

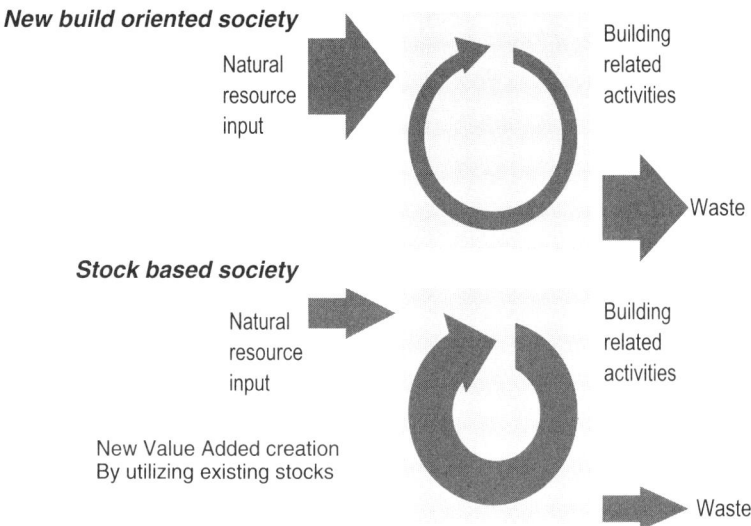

Fig. 2-12. Idea of open building

Inspired by the idea of open building, the financial/tenure system of the houses is proposed by Mr. Nakazato as shown in Fig. 2-13. Here skeleton is owned by a single responsible body and the financial arrangements are on a long-term basis so long-term-based investors such as pension funds could provide the financing. In this case, the skeleton is maintained by the single responsible body and an engineering report is issued to the investors and shareholders. In the diagram shown in Fig. 2-13, the infill is considered as personal, property, so the short-term house loan, for 10 or 15 years, becomes just a small amount.

Thus, by following such a scheme, the value of the skeleton and the value of the infrastructure of the building can be maintained on a long-time basis, while the value of the infill itself can drop down and jump up repetitively on a short-term basis. In this way, 70 percent of the value of the building can be maintained in the long run.

2.3 How Could We Improve the Situation?

2.3.1 Stock Management in an Information-Driven Society

In order to mitigate the serious problems of the "poverty trap," we need to establish and practice a method of stock management in the Japanese con-

Idea of open Building

Skelton/Support

- Use of 100 to 200 years
- Requires continual engineering assessment report based on monitoring

Infill

- Private property exchangeable respecting on ever changing requirement

Fig. 2-13. A. 11. Housing loan system corresponding to open building idea

text as fast as possible to enhance the value of the existing building stock. It should be noted that a fundamental change in the global socioeconomic and technology systems are underway, especially in the well-developed world; information itself is becoming one of key driving forces in society. It can be said that the society is getting to be information-driven where information itself generates added value.

Various stakeholders are committed to the decision-making process of stock management, and there is a serious need to provide understandable and traceable information with engineering integrity in order to promote continuous investment in existing building stock and facilities.

2.3.2 Information-Embedded Building

Performance and quality are the key factors in identifying the value of existing buildings. However, these factors are invisible and difficult to assess for nonprofessional stakeholders. Therefore, assessment of performance and quality of buildings, and provisions of assessment report to stakeholders is getting to be quite significant. To make the assessment report traceable and reliable, it is mandatory that the assessment is done by evidence basis.

In reality, not all buildings have sufficient evidence usable for assessment. Information relating to a building is generated by many agents such as the architect, engineer, main contractor, subcontractors, suppliers, operators, maintenance teams, owners, occupants and users, etc. Generally, information generated by some agent is saved and used by that agent only.

Therefore, from a bird's-eye point of view, a set of information needed for a comprehensive assessment of the performance and quality of a building is fragmentally maintained and used by various agents. Thus, it is difficult for a stakeholder to collect necessary information at any specific occasion because of fragmented information management. In a sense, a building is a black box for stakeholders. No one can put value into or make an investment on an object whose performance and quality are unknown. This is one of the reasons why the value of Japanese buildings follows a conventional evaluation curve over time. In other words, if we intend to enhance the value by jumping up from the conventional curve, we need to assure that full accessibility to a set of comprehensive information is available anytime for any stakeholder who has socially agreed permission to access the information. When reliable, traceable, transparent information is provided, the value could be increased as seen in Fig. 2-14. And if the content of provided information is of high quality, proving that this is a very high-performing building, its value could be additionally enhanced as seen in Fig. 2-14. We can thus expect an enhancement in value by making the building traceable and by providing evidence or excellency when the building displays good performance and quality.

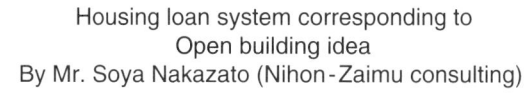

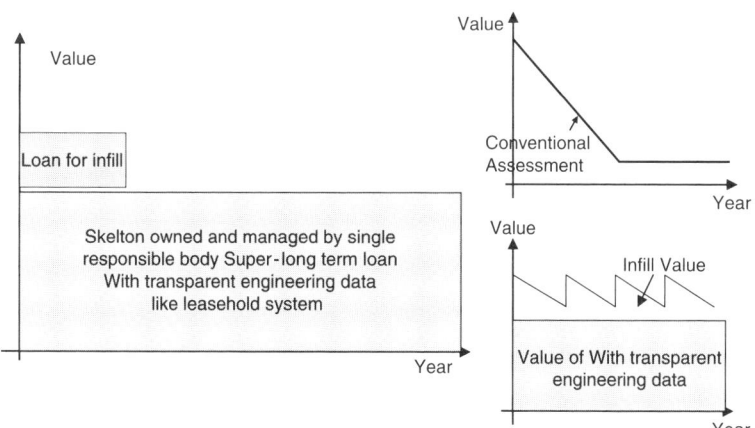

Fig. 2-14. Enhancement of value by information provision
Significance of provision of traceable and understandable information with engineering integrity

Here the idea of an information-embedded building is posited to assure full accessibility to a set of comprehensive information (Fig. 2-15). It is a building for which on-site accessibility to necessary information can be provided, as supported by ubiquitous computing and information networking technologies.

In an information-embedded building, for instance, structural performance and energy use are monitored using networking technology as well as innovative sensing devices. In addition, the history of the maintenance of components or equipment of the building can be traced using IT tools. The following technological developments by the author are examples of subsystems that support an information-embedded building

2.3.3 Life-Cycle Traceability System

The first example is a life-cycle traceability system of building components and equipment using RFID (Fig. 2-16). Regrettably, the repair and mainte-

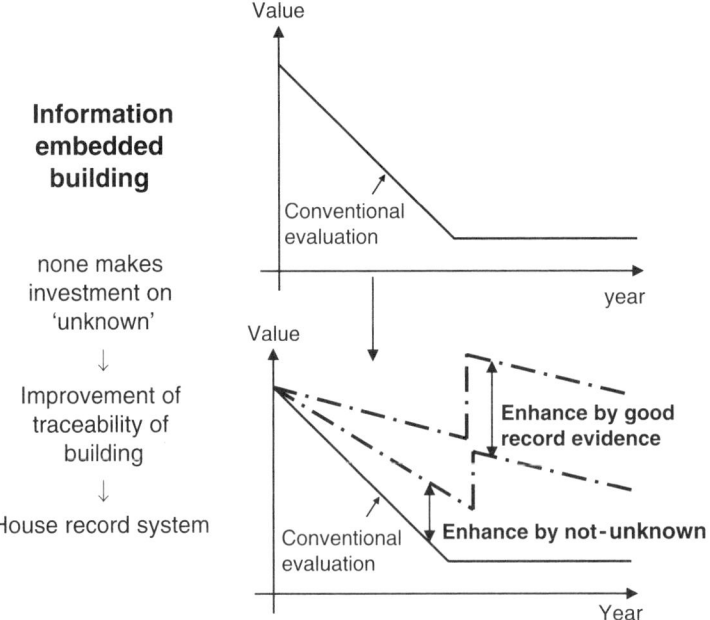

Fig. 2-15. The idea of an information-embedded building
In an information-embedded building, professional and nonprofessional stakeholders have access to information for well-informed decision making using IT devices (including RFID and barcode, sensors which are connected, systematized and integrated by Internet/Intranet)

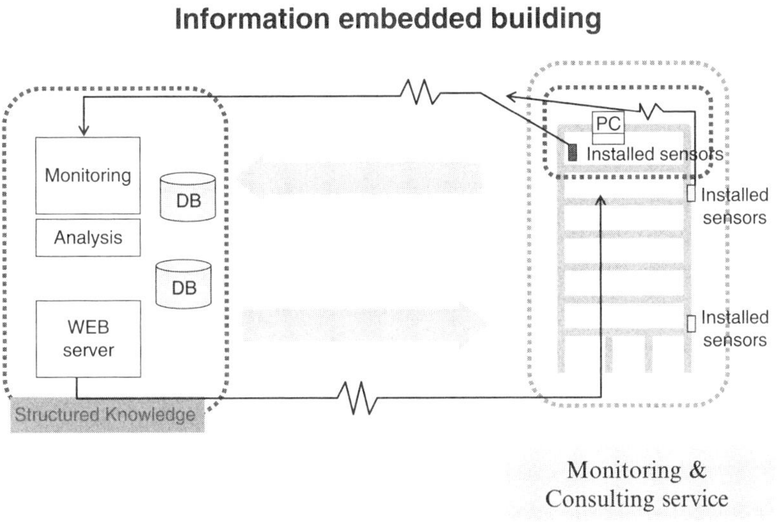

Information embedded building

Fig. 2-16. Life-cycle traceability system using RFID

nance of components and equipment are generally inefficient because site operatives have difficulty in accessing the necessary information for repair maintenance. It is a time- consuming task. Site operatives often take longer time to access, seek, collect, send and return information than in the actual repair and maintenance work. Moreover, there is no assurance that maintenance record data are stocked and used consistently. In reality, the loss of maintenance record data degrades the efficiency more seriously.

As Fig. 2-16 shows, a life cycle traceability system using RFID improves the inefficiency of repair and maintenance and loss of maintenance record data: site operatives of repair and maintenance can easily access the information needed for execution of onsite tasks such as previous maintenance record, technical specification of components and equipment, and detailed technical information about the fabric of the building. The repair and maintenance record data are sent to a database almost automatically and a revised data base is updated consistently for future repair needs.

The traceability system reduces the risk of the cost increase of maintenance, repair or refurbishment and provides benefits such as:

- Faster response to technical failures
- Proactive user support by suppliers
- Links with electronic manifesto certificate for waste treatment

2.3.4 Energy-Use Monitoring System

Fig. 2-17 shows an example of the output of a real-time-based energy-use-monitoring system developed by the author and his colleagues. Energy-use data can be collected through the intranet network installed in the building, and the results of structural analysis are displayed according to the needs of the users. Through the monitoring system, all stakeholders of the building can see how much energy is used in each part of the building (of course there is a need for access control to protect privacy). This system could be a mechanism for promoting voluntary actions by various stakeholders of the building to improve the energy use, because anyone can see and learn how his/her actions contributed to the end result.

2.3.5 House Record System

The author and his colleagues also developed a house record system where a set of drawing specifications, maintenance records, and energy use data is maintained. Fig. 2-18 is the example of the display of this house record system. In a sense, the system is the view window of a set of comprehensive information used for evidence of assessment of performance and building quality.

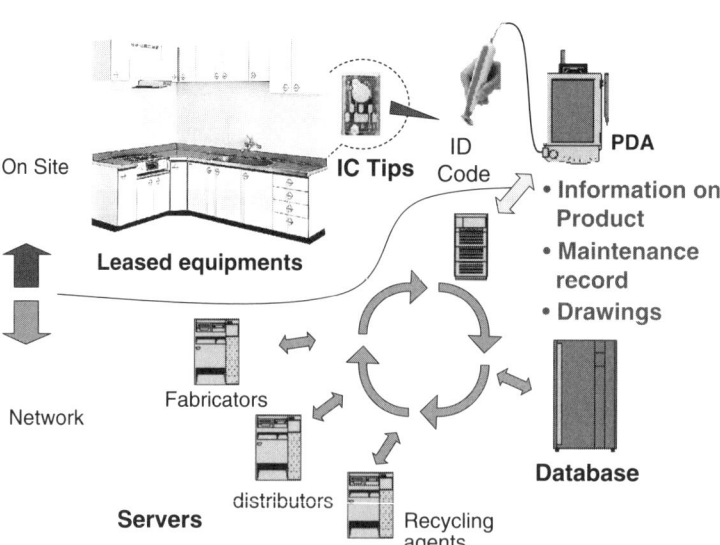

Fig. 2-17. Real-time energy monitoring system

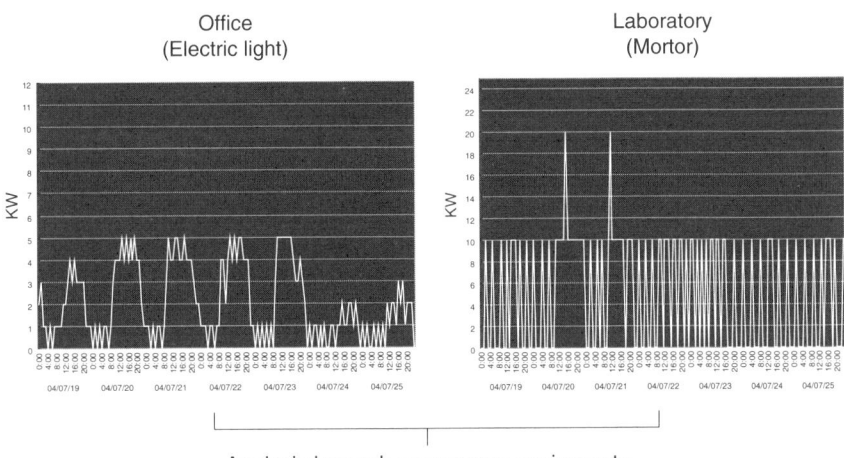

Analysis by each energy use equipments

Fig. 2-18. House history system

2.3.6 Implication of an Information-Embedded Building for Stock Management

Then, what are the implications of an information-embedded building for stock management?

In a time of structural change in a construction market dealing with modifications to existing buildings, as well as in a time when an information-driven society is emerging—one where many stakeholders can participate in and commit to the process of decision making about stock management—the building industry itself needs to be fundamentally changed.

It should be noted that, when compared with new building activity, actual stock management contains uncertainty because of the lack of information. Also, currently it is difficult to introduce the benefits of economies of scale because most stock management requires highly individual, very specific components and materials. The manager of the stock management project is required to have patient dialogues with various stakeholders; linkage to the financial sector is required as well. Information-embedded buildings provide traceable information usable for this multi-stakeholder dialogue.

Thus a completely different knowledge is required for new building projects, one that is quite challenging. There is a need to create a body of empirical knowledge first of all through experiments; this empirical knowledge then has to be systematized. The information-embedded building is the device that will enable professionals to build up empirical knowledge of stock management with certain engineering data. This is a challenging area for younger professional and students, who are expected to do make their contributions toward resolving these difficulties and challenges.

2.4 Concluding Comments

For stock management of urban facilities, a holistic approach is needed. "Holistic" means that one can collaborate or team with people having different expertise—not only building engineers, but also financial people, local stakeholders, and so on. In the existing academic domain, that could either be a constraint or an obstruction, because the issues at stake today are completely different from those derived from academic traditions as described here. Probably most readers have dealt with these issues where problems are simply defined, where the number of decision makers is limited, and where there is no discussions about norms. In this scenario, a simple principle can often be applicable to describe the whole system. But in terms of stock management, none of those traditional typical attitudes are valid any longer; in other words, some really complicated matters now arise: the number of decision makers can be unlimited sometimes, or the discussion can sometimes extend to the field of norms and ethics. Also there are new issues around the parameters of what is being created: various principles can describe a part of the problem, but no single principle can describes the whole system. This is the area of challenge, practically or academically.

Thus, the author charges the student: "Be ambitious, everybody, in defining the next generation of practices and to work on them beyond the existing academic domains."

References

Meikle, J.L., Connaughton, J.N. (1994) "How long should housing last? Some implications of the age and probable life of housing in England", *Construction Management and Economics*, 12(4):315–321

3. Renovation of Modern Stock

Yukio Nishimura

The objective of this paper is to show:

1. How to manage history in an urban setting
2. Urban heritages in the wider context—not only in terms of individual buildings, but also in the context of urban tissue, infrastructure, and street patterns and
3. How to understand, appreciate, and deal with these urban heritages

Sometimes we have to change important urban stock in redesigning our city; in order to do that without any harm to the surroundings in consideration of the urban complex and its history, we need to understand the real meaning of our urban heritages.

3.1 Lesson from the University of Tokyo and Its Surroundings

3.1.1 Overview of the Past UT

In this historical map of 1883 showing the area around the campus of the University of Tokyo (Fig. 3-1), dark color indicates the brick walls and other fireproof buildings. At that time there was a long main entrance, what we call a *seimon* (the main gate), and it was located as you can see in the picture. The main entrance of that time is different from that of the present. Also, the original campus was where the medical college is located now. Main access of the road at that time was separating the medical school

Y. Fujino, T. Noguchi (eds.) *Stock Management for Sustainable Urban Regeneration*,
© 2009 to the complete printed work by Springer, except as noted. Individual authors
or their assignees retain rights to their respective contributions; reproduced by permission.

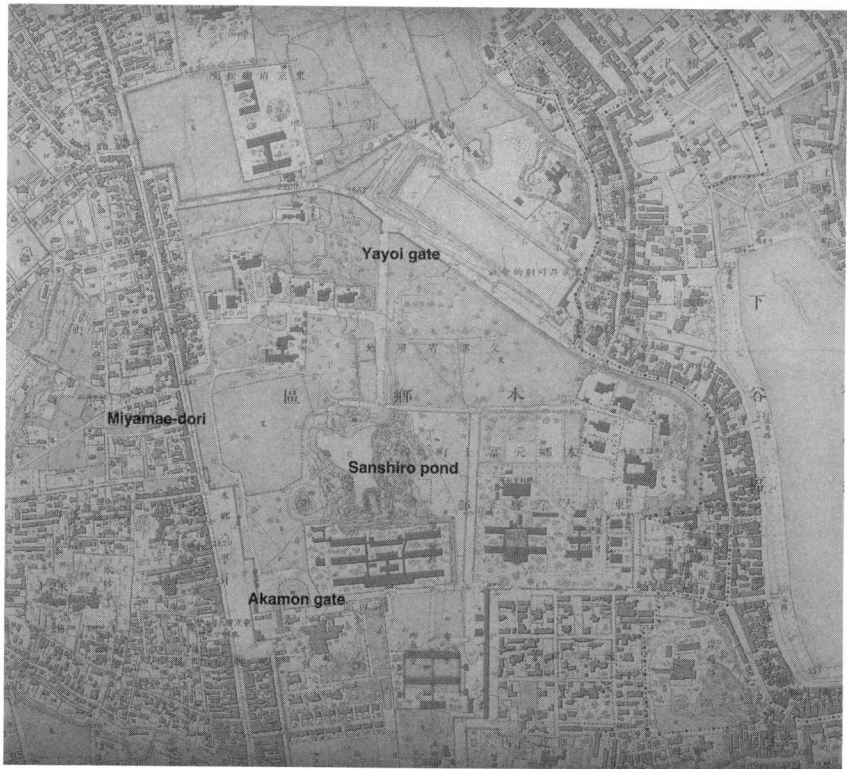

Fig. 3-1. 1883 Map around University of Tokyo by the Bureau of Cartography, the Military of Japan (scale 1:5,000), top being the north

and the rest of the compound. Also, the Sanshiro pond used to be a private garden of the owner's residence.

3.1.2 Process of Formation

The whole compound was owned by a single family, which was a manor family of Kaga-han, which is Ishikawa and Toyama prefectures today. As you know, in medieval Japan, all the landlords of manors in the rural area were requested to have their own Edo, former Tokyo residence, some-where very close to the capital. So there were main Tokyo residences and the every year or every other year, depending on the distance from Tokyo, the lords of the manors, called *daimyo*, were required to come to stay in Tokyo for one year and return to their own country the next. Consequently, they were all commuting every year or every other year with hundreds or sometimes thousands of warriors.

This was therefore a manor house, but at the same time, an accommodation of a large number of the warriors who supported the manor family. It meant that all the accompanying warriors had a chance to study, to collect many things and information, and to get accustomed to the high culture at the capital. This is one of the unacknowledged reasons that Japanese modernization was so quick.

After the Meiji reform in 1868, however, the warrior class was abolished. They had to leave Tokyo and settle down in their own country and sometimes set up commercial activities. It meant that many huge lands became vacant at that time, so the central government confiscated these lands to convert the sites into modern facilities like the City Hall, government buildings, universities, and many other public edifices. This is how a vacant land owned by a single family turned into the first national university of this country.

3.1.3 Structure of the Campus of the University of Tokyo and the Surroundings

3.1.3.1 Inside the Campus

It is clearly seen in the picture (Fig. 3-1) that all the development of this area was from the south. From the Tatsuokamon gate, which is on the very old axis, turning right to left, and the Akamon gate, one can recognize the oldest line of trees in this area. All the buildings were demolished mainly by the earthquake in 1923, but were eventually renovated and extended to the north. The guest house of the university used to be the residence for invited foreign scholars.

3.1.3.2 Surroundings—Nezu District

The surroundings have changed slightly over the years. For example, the road Hongo-dori and the Akamon gate were moved slightly, enlarged, and networked. Today we have a very wide slope down to Nezu, and the Kototoi-dori and another road, Shinabazu-dori running through this area. When one leaves by the Yayoi gate, there used to be a firing range for shooting practice. It is strongly recommend to walk this original, historic road because many old temples are located along the way.

3.1.3.3 Surroundings—Hongo District

When one looks at the other side of the campus, Hongo district, there is a whole built-up area. As seen in the late eighteenth-century map (Fig. 3-2), land use was clearly demarcated, pink being for temples, purple for lands owned by the temple, and grey for mercantile establishments. There were the warrior

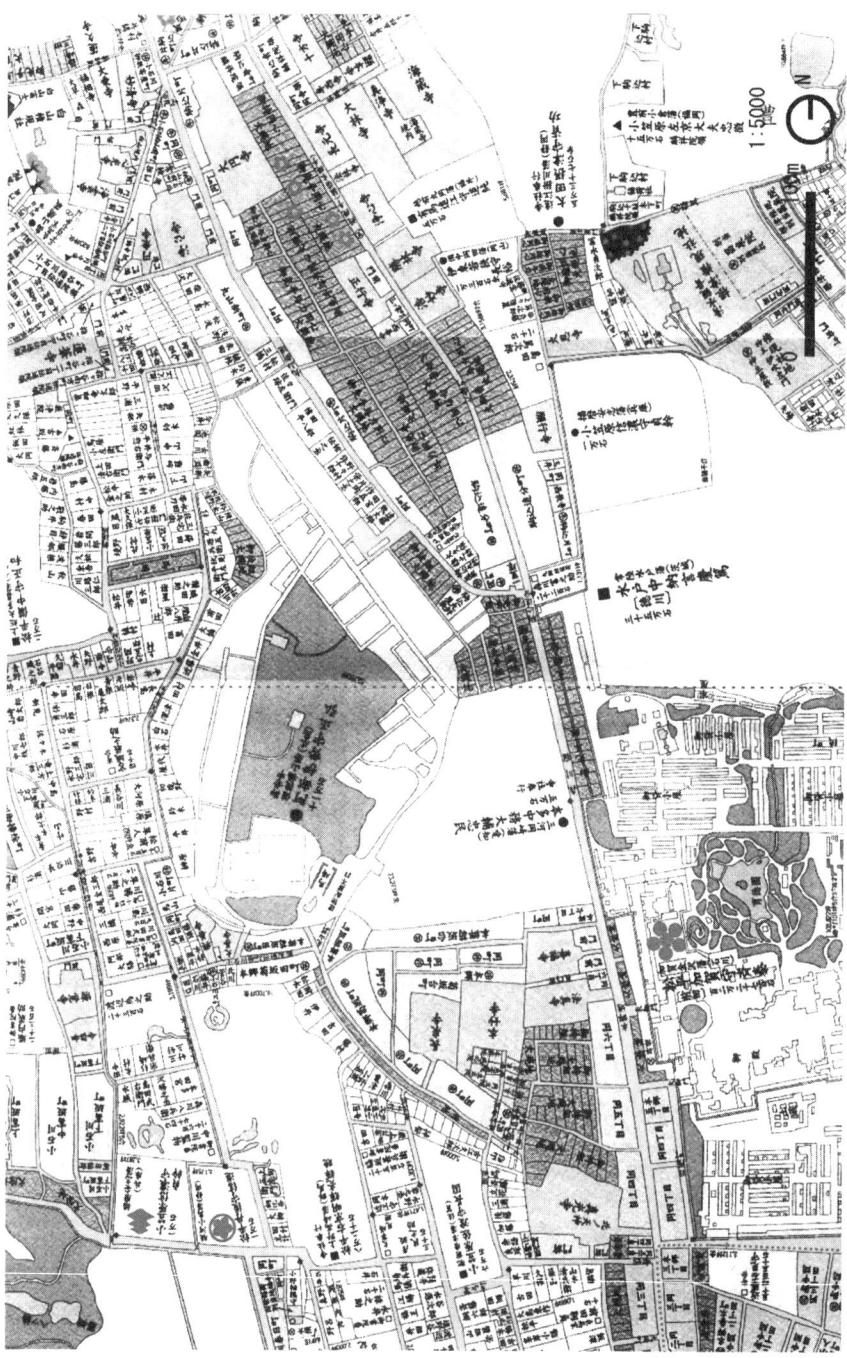

Fig. 3-2. Late eighteenth century Map of the west side of the university campus

class residences and the commercial residences with a temple in the middle, so the respective street plans were completely different from each other.

Going out from the main gate, and turning right and left, one can see a small street, Miyamae-dori (Fig. 3-3). The four pictures of the Miyamae-dori shows this small lane as a main access to the former shrine site, which was located at the end of the lane and later abolished and subdivided into anonymous townhouse lots. However, the name of the lane, Miyamae-dori, literally Shrine Front Street, and the rather compact and straight townscape with the lost shrine site at the end, is almost forgotten, barely indicates that this small lane was designed as an access to the shrine.

3.1.3.4 Surroundings—Focal Point

There is a very strange focal point at the end of the Miyamae-dori. Japanese houses are usually made of wood, which means you need to create a building by posts and beams on a rectangular plot. This rectangular plot is convenient to build, and it is also the traditional type of Japanese buildings. So the rectangular street pattern is easy to understand. However here, this focal point is very strange because it has a sharp angle so that buildings are too difficult to fit in the rectangular plot. Why was this kind of pattern created in the late nineteenth century?

Fig. 3-3c shows a triangular space in the middle of the site. When one looks at this map again, it can be seen this area used to be a shrine, called Eisei shrine, within a huge compound of warriors' residences. It was decided to make use of a small private shrine in this area and make approaches to the shrine with shop houses, which enabled the builders to subdivide and develop the area as residential and commercial quarters. They tried to make a good focus for this area by this small development. That was the role of the shrine. This small space (focal point) was the open space in front of the shrine. But now, the shrine is already gone and all has been changed.

Fig. 3-4a is a famous picture of the small focal point in the 1920s taken by Ihei Kimura, one of the greatest photographers in modern Japan. This shows the focal point of different activities, at the different angles of the roads. The former police office here has been converted into some kind of firefighting activity center (Fig. 3-4b).

The map in Fig. 3-5 was created by our lab members, based on several historical maps of different periods. The colors indicate the different time spans in which the road was constructed. It also shows the analysis by our lab members. How had this strange junction been created? Our hypothesis is that this axis came from other directions crossing over the

Fig. 3-3. (a–d) Miyamae dori, a small historic lane leading to the former shrine site at the end

a

b

Fig. 3-4. (a and **b**) **(a)** Photo of Ihei Kimura in the 1920s featuring the Miyamae-dori area, **(b)** the same site today

small valley, which created another junction on another road, and so on. Because this area was also owned by a single family and (re)developed sequentially from the north to the south in the late nineteenth century, it was rectangular.

Roads were developed and connected to this area. Crossing a bridge connected to another planned area to the northwest, one can see that the plot is quite different, with much bigger detached houses. These are houses of high quality because the owner of this area tried to create a high-quality residence catering to the teaching staffs of this university. Therefore, when one walks around this area, one can feel the townscape is quite different.

Fig. 3-5. Street pattern of the Hongo District, showing the evolution of the blocks, Darker colors indicating the older streets

At the southern part of area shown in Fig. 3-2, in the middle, there used to be a temple surrounded by a mercantile settlement. The street pattern is exactly the same as that in premodern times, even though it might have had a different name and a different landowner.

In the late nineteenth century, this old street pattern was quite different from that of the present. Today, if one walks near the post office on the Hongo- dori, one can notice that there is some gap between the southern part and the northern part. This gap is also clear with the name of this district; Hongo 5-chome and 6-chome, each of which used to be a warriors' holding and a merchants', respectively. This gap therefore came from the two different old land uses. Also, there is no straight road crossing this area. That is because the consequences of development were quite different from each other.

To sum up, when one understands the history and subsurface design concepts of an area, one can have a clear understanding of that area. Also, when architects or designers today cope with the area as well as individual buildings, they can have a clear idea of what should be done or what should not be done to make visitors easy to appreciate the area. The first thing they have to do is to understand the area.

3.2 Message from Ueno Park

3.2.1 Introduction

Ueno Park used to be a huge compound of temples. The whole area, including the Tokyo University of Fine Art, the cemetery area, and the shop houses area near Kototoi Street, was all owned by a temple called Kanyeiji. This huge temple compound was created in 1625 by Shogun Tokugawa Iemitsu to house cemeteries of the Tokugawas and at the same time to protect the northeast part of Edo. It was believed that the northeast part of Edo was very vulnerable to the devil, so they created a huge temple in the northeast side of the capital to protect the residents. Fig. 3-6a shows the land use of the site in the eighteenth century, as superimposed on the current map.

At the eastern part of Ueno Station, there is a commercial street whose networks were made in the medieval era. But the western part of Ueno Station is quite different. Current visual differences are based on historical land uses.

a   b

Fig. 3-6. (a and **b) (a)** Shows eighteenth century land use on top of the current map, while **(b)** illustrates the early stage of Ueno Park in the late nineteenth century

Here is another picture of the area in the late nineteenth century (Fig. 3-6b). It illustrates the early stage of Ueno Park, which had been converted into that use in 1883 after all the buildings were burned by the civil war in 1868.

3.2.2 Design Concept of Ueno Park in the Past

Ueno Park used to be a part of the compounds of the Buddhist Kanyeiji Temple, and the central government tried to reserve the spatial structure of this temple compound, including the main hall of the entrance. They appropriated the site and designed a modern-type park, keeping the former spatial structure of the temple. Its spatial structure can thus be appreciated: for example, the main access should be from the south to the north, or the layout of the exposition held at this site should be symmetrical within this axis (Fig. 3-7).

But at the same time, the central government tried to deny or neglect the shogunate legacy by replacing the memory of that once-great temple by then ultramodern urban facilities such as museums and expositions.

In the late nineteenth century, several expositions were held at this site, whose main access is very similar to previous access routes to the Kanyeiji Temple. From 1911 to 1953, there were several new buildings created along the main axis (Fig. 3-8).

3.2.3 The Changing Ueno Station

Today a big elevated highway runs in front of this station, so it is regrettably impossible to see the kind of view as shown in this Fig. 3-9. The Ueno station was renovated in 2002 to accommodate fashionable restaurants and boutiques around and above the main concourse. This picture shows the old main entrance, which has greatly changed its appearances. Only the huge and high ceilings at the former main entrance section remain to give a good sense of the past. By going to the second floor of Ueno station, one can see how this old station was renovated to modern use.

3.2.4 The Development of Ueno Park

The Park was renovated block by block in the 1910s. There was a series of extensions of the National Museum from 1932 to 1956, and then another access to the Park was gradually created to cross the main access, followed by the introduction of the Ueno Zoo, the Ueno Public Hall, and the

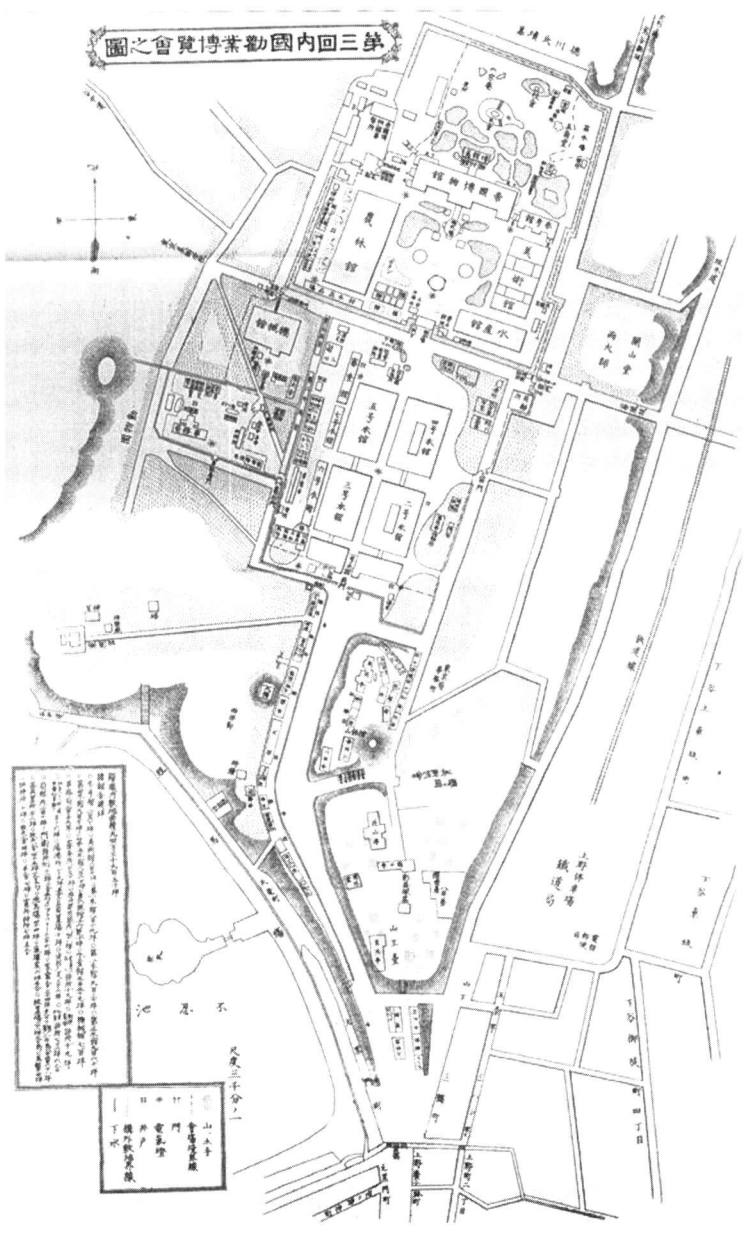

Fig. 3-7. The Map showing the layout of the 3rd National Industrial Exposition held at Ueno Park in 1890

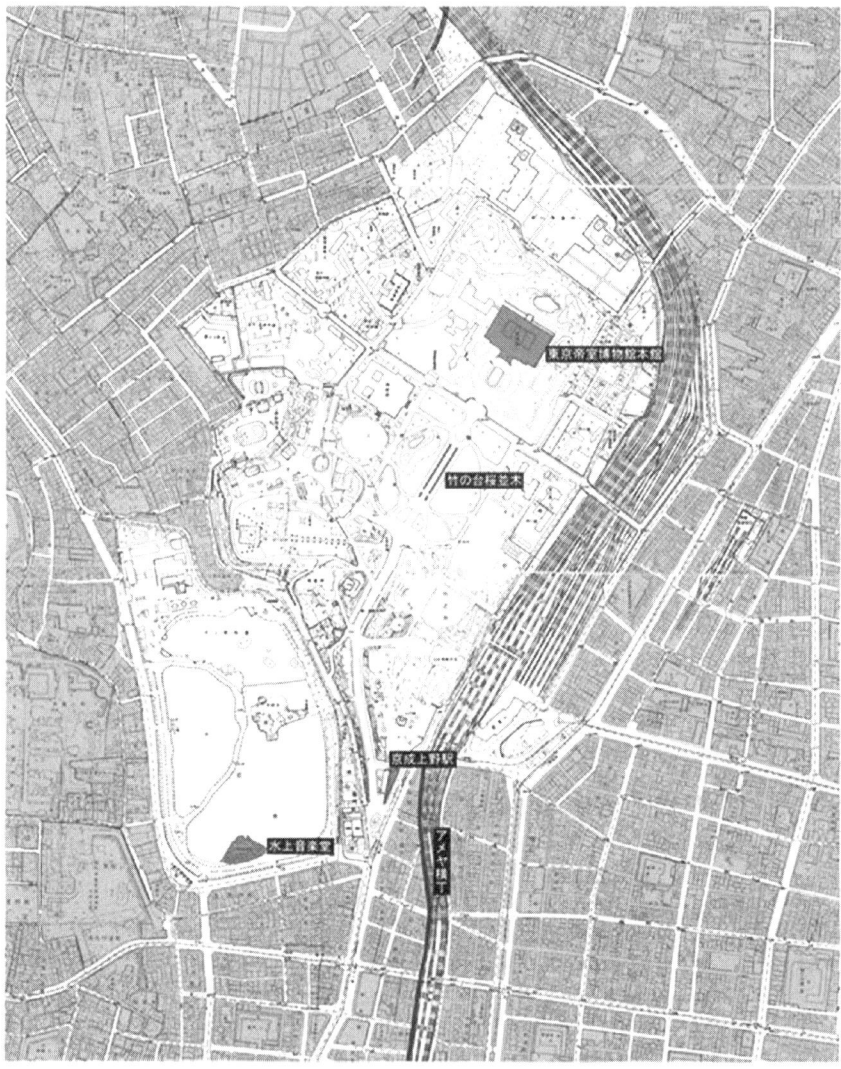

Fig. 3-8. Layout of the Park in 1953

Monorail in the Zoo. As of 1983, the main and secondary access from the station to the entrance of the Zoo formed a cross shape in the area.

3.2.5 An Experiment and What Was Learned from It

In the case of Ueno Park, for example, the main axis is the key to understanding the spatial structure of the site, which should be conserved. There

Fig. 3-9. Façade of the Ueno Station when it was reconstructed in 1932 after the 1923 Kanto Earthquake

is an outdated, huge spring designed in 1962, which is isolated from the people by a huge hedge. More interaction between the water and the people is needed here, so I strongly recommended redesigning this area in order to make an intimate open space for modern activity by reinforcing the main access to the Tokyo National Museum by the Tokyo Metropolitan Government. However, first of all, the spatial structure has to be understood, and its importance.

In front of the Tokyo National Museum, which is the main focal point of the park, there is a vacant space. According to our evaluation, it is not good to have this kind of huge vacant land at the crossing center of the two axes. People need temporary resting places and kiosks or something removable, and also need information about the different museums, galleries, etc. These kinds of activities are strongly needed here.

Therefore, several years ago, we proposed an experimental kiosk and coffee shop on the axis in front of the National Museum where anonymous vacant space had been created. Fig. 3-10 shows the site before and in the middle of the experiment. It lasted for only three to four weeks, but it fostered good activities, and also made a good amount of money by the sales of beverages. The income can and should be returned to maintain the surroundings.

However, according to the general rules of the Japanese administration, it is usually difficult to make profits generated from the public space, because it is very difficult to decide who is responsible for the monetary activities. The

a

b

Fig. 3-10. (a and **b**) Vacant 'public' space in front of the Tokyo National Museum (**a**) and experimental café and kiosk at that site (**b**)

Fig. 3-11. Statue of Dr. A. F. Baudwin in Ueno Park

project was very successful, but unfortunately the local government did not permit the additional facilities and activities any more, because no one really took the risk and expended the effort to continue this project. Regrettably, it is safer for a typical government official not to commit to any proposed changes because it may cause some unforeseen problems. And since it is a public venue, no one could contravene this common attitude of a common officer. As a result, no one is responsible, and public space does not become everyone's land but no-man's land. Thus, no lively activities can emerge from such "public" space, to say nothing of income generated there. We need to rethink what is 'public' stock after all, and who is responsible for this space.

3.2.6 Statute of Dr. A. F. Baudwin

Dr. A. F. Baudwin (Fig. 3-11), a Dutch medical doctor, proposed this area as a park in 1870, when he was invited to give some advice about the former temple compound. He thought that a park was much more suitable for this area than the hospital that had been originally proposed by the Meiji Government, because he could read the context of this spacious and sunny area. He suggested that hospitals could be created anyplace, but it was very difficult to obtain such a huge tract of land that was sunny, clean, and easily

Fig. 3-12. (a and **b)** Viewing terrace of the Kiyomizu Temple in the early nineteenth century from the wood print (**a**) and current viewing terrace (**b**)

accessible for the general public. Thus, a modern park, Ueno Park, was born in 1873 for the first time in Japan's history. It was one of the five modern parks that were decreed at the same time by the central government.

Unfortunately, however, this important history seems to have been forgotten, and the statue of this important person is regrettably now in a very bad situation surrounded by the homeless. It is important to recreate a beautiful, small square as the focal point of the main access to memorialize his work.

3.2.7 Viewing Terrace of Kiyomizu Temple

As can be seen in the two pictures of the past (a) and present (b) (Fig. 3-12), there was once a viewing terrace here at Kiyomizu temple, designed for looking

Fig. 3-13. Hiroshige's drawing of the viewing terrace of the Kiyomizu Temple

out over the Shinobazu Pond and farther west. However, the view today is entirely blocked by cherry trees that were planted there after World War II.

Ando Hiroshige, the famous ukiyo-e artist of the early nineteenth century, depicted from this terrace the same view of the pond through a twisted pine tree (Fig. 3-13). Many artists drew pictures of this view, which was quite typical of this area.

Also, the Kiyomizu Temple has been designated as an important cultural property by the central government. Fig. 3-14 shows the current view of the terrace. Is this view acceptable with the historic vista blocked? Not at all. But when one negotiates with the proprietors, it is very difficult to convince them that the vista is more important than branches of cherry trees.

Moreover, almost half of the Pond now belongs to the Ueno Zoo, which has restricted access to the site, which means people cannot go around the Pond to make a circuit. But as can be seen from the historic drawings, the

Fig. 3-14. Cherry tress blocking the view from the terrace today

Fig. 3-15. Early nineteenth century drawing of the Shinobazu Pond

original idea of the Pond included to be a promenade where visitors could circumnavigate the whole Pond.

When one looks at these spectacular historical drawings (Fig. 3-15 and Fig. 3-16), one can realize how actively the promenade had been used in the past. For example, Fig. 3-16 shows horses racing around the Pond after a racetrack was introduced in 1884. This is one of the birthplaces of modern

Fig. 3-16. Late nineteenth century drawings of the Shinobazu Pond, depicting the horse racing

horse racing in this country. All the audiences were then sitting along the Pond. On other occasions, this area had been converted into a temporary pavilion for exhibitions. One can imagine how closely interactive activities existed along the Pond.

Similar activities should be introduced to revitalize these lively scenes from the past—not necessarily the exact same things. This was the main purpose when the Pond was converted into a part of the Ueno Park. Planners should try to extract ideas from the past for current possibilities or activities and then suggest them to those concerned.

3.2.8 Access to the Pond

These pictures (Fig. 3-17) show the promenade of the Pond, a part of which has been closed by the zoo, resulting in no activities because of no entrance. There is no reason for going there, so it is difficult to recall or recognize what the place is like. There are no activities, nothing to do, with only some homeless people are staying around the promenade.

From the historical pictures, one can interpret the possibility or the potential of this area to reintroduce and rearrange much closer interaction with the Pond. One can visualize the future possibilities of the site by appreciating the historical activities.

As seen in Fig. 3-18, these kinds of high-rise buildings unsympathetic to their surroundings are being built today. They should be stopped. However, our Building Codes are somehow inadequate to control this kind of con-

a

b

Fig. 3-17. (a and **b)** Promenade of the Shinobazu Pond today, frequently used by the public in the southern part **(a)** and shuttered by the exit gate of the Ueno Zoo in the northern part **(b)**

struction because it is controlled by the width of the adjacent road. If there is an open space just close to the construction site, then one can design higher buildings, because no one is living in the adjacent open space.

This is ridiculous because everyone using the Park and the Pond enjoys views from the Park. High rise flats along the Pond damage the panoramic view of the Park. Therefore, it is necessary to change this kind of incentive planning system, which creates extra pressures on public spaces. We should

Fig. 3-18. High-rise buildings unsympathetic to the surrounding environment are being built around the Shinobazu Pond

seriously contemplate what "public" is, what the use of "public" means, and what is good for the general "public."

3.2.9 Conclusion: On the Ueno Slope

Fig. 3-19a shows another main entrance to Ueno Park, on the southeastern side. It was beautifully designed in the late 1920s after the Great Kanto earthquake in 1923. Thanks to the finely landscaped garden, the Park attracted a large number of people as depicted in the postcard (Fig. 3-19a).

Shamefully, the place today is quite ugly, as seen in Fig. 3-19b. The famous Saigo Takamori's statue has been visually damaged after World War II because of the huge squat of small shops that had been created after the war. Tokyo Metropolitan Government tried to remove the squatters, but usually it cannot be done without just compensation, and if the squatters settled for a certain period of time, they eventually acquired the right to claim the compensation no matter how the ownership is. Finding it difficult to make additional space where the squatters could engage in retail activities, and faced with soaring land prices, the government finally decided to make a huge hall at the foot of this slope to accommodate these illegal squatters.

a

b

Fig. 3-19. (**a** and **b**) Prewar postcard depicting the large crowds in the Park (**a**) and the same site today (**b**)

This seemed to be the easiest solution to the problem. Now when one look at the Shinobazu entrance to Ueno Station, one cannot fail to find this unsightly building that cannot be recovered even today.

Therefore, the mission of architects and planners is very clear. In this case, it is obvious that we should recover the original grand slope of the Park. The façade of the Slope site should be public.

This kind of idea is important to restore or reuse historic infrastructures as well as historical street patterns. From what has been seen from the examples close to our campus, we should consider what we should do for the public, what the main clue for a future vision is, how to maintain and renovale our urban stock, and how to conserve our heritage for future generations.

As a planner dedicated to urban conservation, this is what I am planning and proposing better urban spaces through appreciating the past for the benefit of the general public of today and tomorrow.

4. Why and How We Should Inherit Urban Environmental Cultural Resources: Identifying, Listing, Evaluating, and Making Good Use of Urban Environmental Cultural Resources in Asia

Shin Muramatsu

4.1 Introduction

A city as a spatial expanse includes people as well as different natural and artificial elements. Let us regard the entirety of such a city as "the urban environment": a stage for human lives, including its space and all the elements within. To classify these various elements is basic to considering and preserving the urban environment, though there are different ways of classification depending on one's purpose. If one intends to build, for example, such elements should be considered as civil engineering works, architecture, other artificial urban constituents, and nature, which correspond respectively to the academic fields of civil engineering, architecture, urban engineering, and landscape design. Such classification, however, cannot cover other inherent urban elements such as air and water, and wastes that are discharged by our urban activities.

Therefore, for the purpose of this program that intends to cover all the elements within the urban environment and to manage their sustainable development, it should better suit us to adopt a new viewpoint in which these elements can be regarded as usable "resources," leaving aside the established academic structure.

Now we can divide all these elements within the urban environment as natural resources or cultural resources. While there have been many discussions regarding "culture" and "resources," here we define "culture" as the entirety of people's living styles and customs that are utilized in the contexts of (1) everyday living; (2) society at large; and (3) the marketplace. Natural resources, on the other hand, refer to elements such as air, water, earth, trees, and living creatures, but they can be also constituted as cultural resources depending on how we treat them.

So far, people have considered the urban environment to be static, consisting of either natural or cultural resources. This way of analysis itself, however, cannot take advantage of the dynamic meaning of the word "resources." A city itself is also a dynamic system that contains human activities, accepting energy, people, things, information, and capital that flow in from the outside world while consuming, establishing, and emitting all those elements, together with the natural and cultural resources within?(Fig. 4-1). Such a system is built upon the sense of values of the inhabitants who live in a particular city. Thus, it is necessary for us to manage the natural and cultural resources that constitute this system called "city" in a sustainable manner so that more people who live and work within it can enjoy its wealth equally from generation to generation.

Particularly in those non-Western countries that were once ruled by colonial powers, people tend to ignore and even reject those things and objects in their living and working environment that have been associated with the colonial past, and such historical circumstances prevent these things from rendering themselves usable as "resources." However, they are indeed the urban environmental cultural resources for those who currently live there. Thus, the authors, based upon their experiences, intend in this article to illuminate the general process of how urban environmental cultural resources

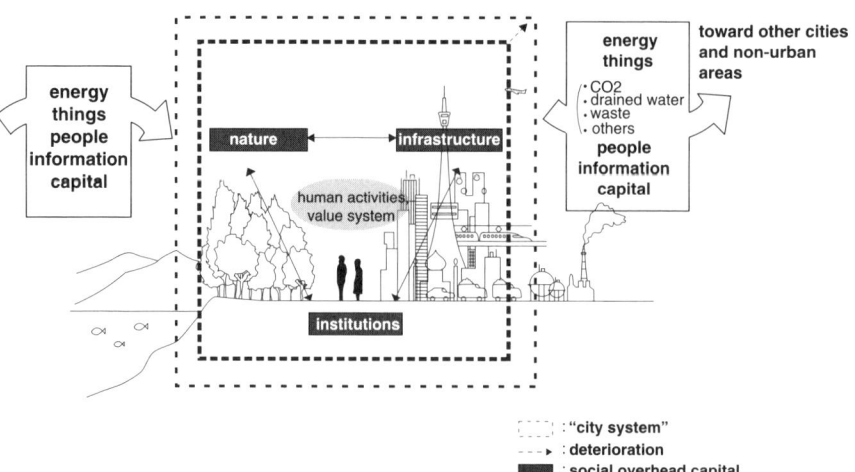

Fig. 4-1. **Temporary** model of city system

could be discovered and conserved in those non-Western cities where such a concept has not been established. This will be done by focusing on their certain important stages.

In the following part, we will (1) define urban environmental cultural resources and clarify the significance of their preservation; (2) explain our methods to evaluate the collected specimens of these urban environmental cultural resources; (3) present our methods for collecting and listing these urban environmental cultural resources, and (4) discuss how we should treat selected resources.

4.1.1 What Are Urban Environmental Cultural Resources?

As stated above, urban environmental cultural resources are the cultural resources that constitute our urban environment as considered together with urban environmental natural resources. Such resources could be utilized for (1) cultural activities realized through our physical bodies (everyday life); (2) socialized cultural activities (social activities); or (3) market transactions (economic activities). The urban environmental natural resources (air, water, etc.) differ from urban environmental cultural resources in that the former are mainly used for (4) human vital activities.

As for urban environmental cultural resources, they may seem similar to the built environment, which consists of urban artificial constructs, but they are not limited to the built environment only and can be classified into the four categories as shown in Table 4-1. Namely, urban environmental cultural resources include not only the things built by people but also any natural and temporal transitions that have been "culturalized." Thus, natural

Table 4-1. Urban environmental cultural resources

Type	Subject	Use
Built environment	Building, infrastructure, artificial landscape	(1) Everyday life
Enculturated nature	Trees, sound, scent, natural landscape	(2) Social life
Enculturated time circulation	Seasonal changes, daily changes	(3) Economics activities
Time process	Impact of irreversible time alteration on the urban environment and on the human memory	

landscapes can be considered to be a part of urban environmental cultural resources. For example, the Eight Views of Xiaoxiang, which were established in a scenic area south of Lake Dongting in Hunan province during the eleventh century (the Northern Sung period), consist of eight separate landscapes each framed by incorporating natural transitions. Later, such "forms" were adopted in various spots of scenic beauty in neighboring countries like Japan, Korea, and Vietnam.

In Japan, also, a book of seasonal words and phrases to be used in composing haiku poems became very popular in the early nineteenth century, and this has carved a strong sense of seasons into peoples' minds. The landscapes thus formalized and the "literaturized" temporal cycles have played a useful function in making people's lives richer. On top of this, people now pay attention to sounds of rain falling and crickets chirping as part of a particular city's characteristics: these are now recognized as some of its urban environmental cultural resources.

4.1.2 Significance of Identifying and Accumulating Urban Environmental Cultural Resources

In this section, we will discuss the methods and significance of understanding, accumulating, and listing urban environmental cultural resources based on our research experiences (Muramatsu (2007)). Urban environmental cultural resources include many different areas, which have been covered by various academic fields. The existing studies, however, tend to present patchy information by listing, in a dispersed manner, examples that were not necessarily selected based on their usefulness for improving the urban environment. Urban environmental cultural resources do not exist as given, but they have to be identified by the people who are involved in each city where they exist.

When we look for and extract specimens of the urban environmental cultural resources in a certain city, we usually follow these five steps: we (1) make necessary preparation for the research; (2) conduct a comprehensive survey of the given city; (3) study relevant documents; (4) try to make sense of collected examples; and (5) extract any urban environmental cultural resources (Fig. 4-2). From our experience, there are at least 100,000 buildings in a city, and we usually cover from 2,000 to 3,000 examples in the course of our comprehensive survey. However, a person who lives in that city cannot easily grasp all these examples as a part of the city's urban environmental cultural resources. Therefore, our purpose of listing these resources is to present the examples that people might not be aware of; they are thus made

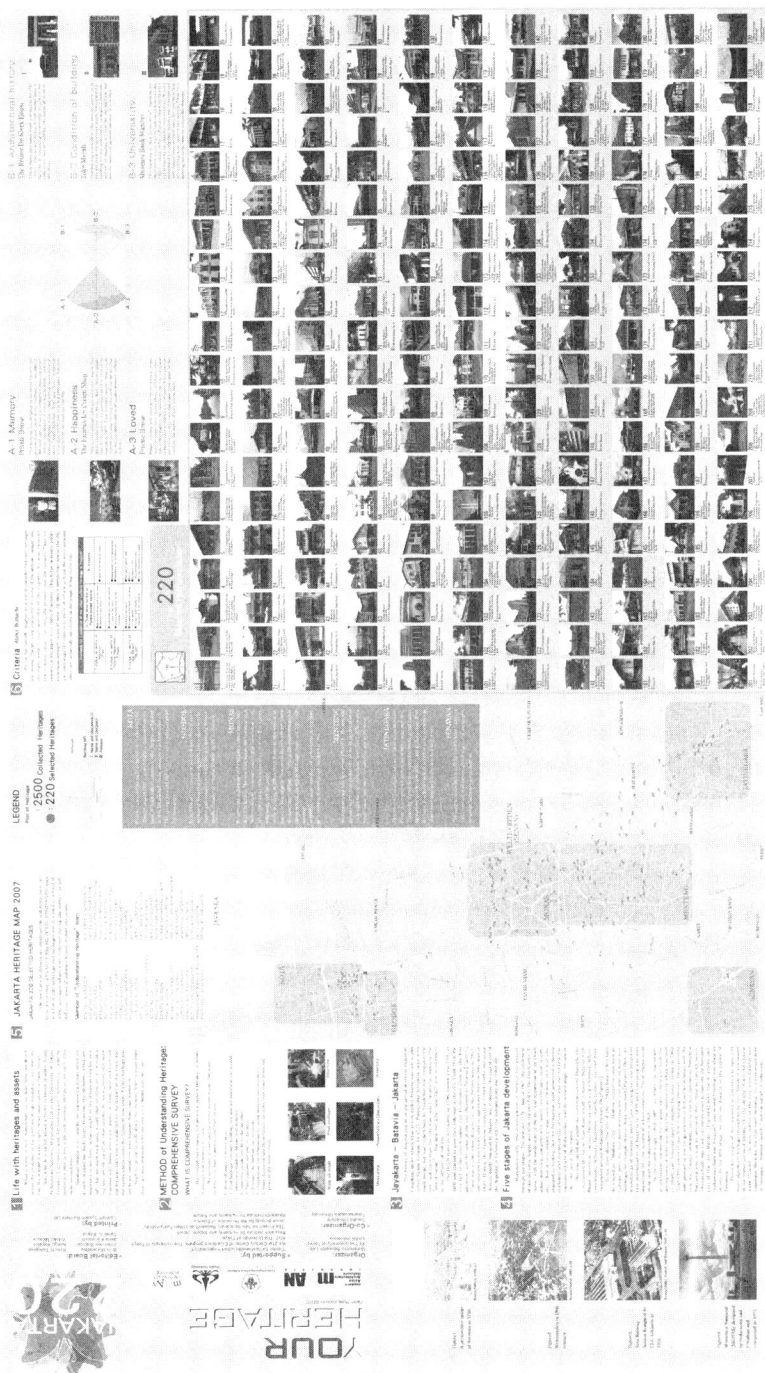

Fig. 4-2. Jakarta heritage map

available as resources that many people can use. For this purpose, it would be a good idea to create an urban environmental cultural resources map on which people can easily see all the examples at once (Fig. 4-2).

4.1.3 Identifying and Evaluating Urban Environmental Cultural Resources: "Heritage Butterfly"

It still remains a problem how to choose from 100,000 buildings and "culturalized" natural resources and to turn what we select into urban environmental cultural resources. In the course of each survey, a researcher is forced to make two decisions. The first decision becomes necessary when he or she conducts an urban comprehensive survey to collect specimens by observing all the buildings in a given city—listed as (2) in the previous section; the second decision needs to be made when he or she extracts what can be constituted as urban environmental cultural resources from the collected specimens. We always try not only to make these decisions from our professional stance but also to reflect the values judged from the users' viewpoint.

Specialists who judge the cultural value of buildings usually do so by focusing on each building's (A) historical aspect; (B) aesthetic aspect; and (C) preserved condition. The basis for such evaluation is found not only within the objects of the survey but also with the researchers themselves. In fact, one of the goals of observing all the built objects in a given city, as in a comprehensive survey, is for researchers to newly build a basis for their judgments. It is also necessary for them to interview the owner(s) and other relevant people regarding each building in order to acquire a non-specialist's sense of their judgments. Such a sense can be based on (a) memory; (b) moving impressions; and (c) maintenance, which respectively correspond to the specialist's judgment bases listed as A–C above.

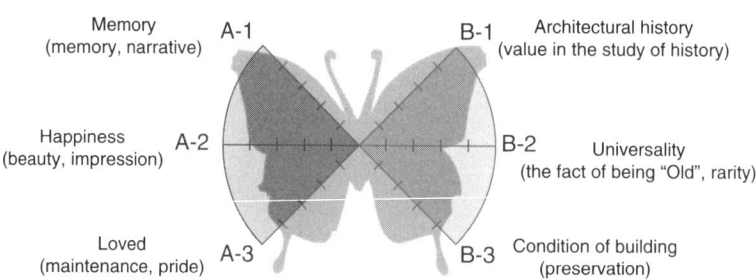

Fig. 4-3. Heritage butterfly

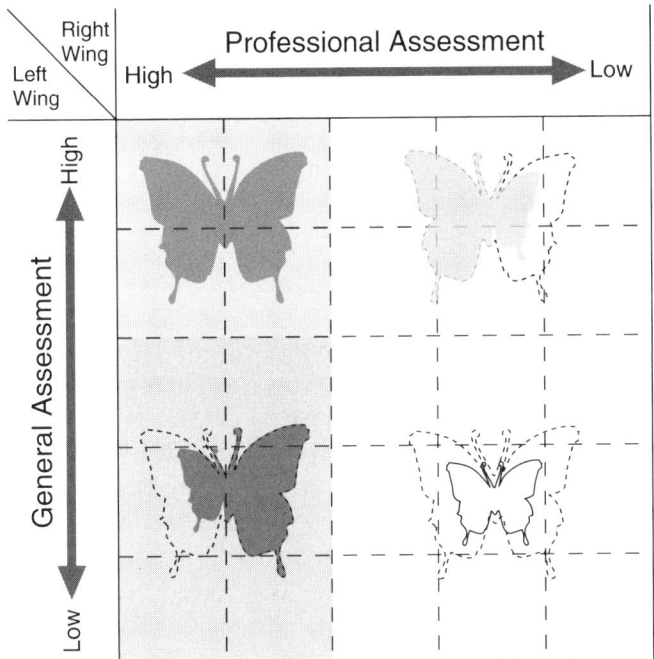

Fig. 4-4. Analysis of activity using 'heritage butterfly' model

What we call the "heritage butterfly" is a graphic device to illustrate these two sets of evaluation bases (Fig. 4-3). This device shows the professional criteria on the right side and the popular ones on the left. A butterfly whose wings are of equal shape and size implies that this resource is of higher validity. However, the heritage butterfly is also an indication of how we should make any interventions, not just the amount of resource value. For example, when the right wing is smaller than the left one (i.e., the professional evaluation is lower than the popular one), it suggests that architects and other professional people should be involved in order to raise the building's professional value. On the other hand, when the right wing is big and the left one small (i.e., the professional evaluation is high while the popular one is low), then grass-roots activists should make people aware of the building's value (Fig. 4-4).

4.1.4 Mechanism for Use of Urban Environmental Cultural Resources and for Literacy Building

As discussed earlier, the best use can be made out of our urban environmental cultural resources in the following three contexts: (1) cultural activities realized through our physical bodies (everyday life); (2) socialized cultural activities (social activities); or (3) market transactions (economic activities). So far, such resources (i.e. cultural properties of different types) have been often utilized in order to raise the profile of a particular community, or to promote tourism. Recently in Japan, for example, people started to pay attention to landscape as a type of urban environmental cultural resources that enriches their lives. But in most cases, urban environmental cultural resources are still appreciated in the sense of (2) and (3). I, however, consider it more important when such resources prove themselves useful in people's daily lives, and the surveys and their results introduced here are also intended for such use more than anything else.

At the same time, it is important to educate young people regarding the significance of urban environmental cultural resources as well as the methods for their comprehension and use. And it is more effective to educate elementary school children whose thinking patterns are more flexible than those of college students and adults whose sense of value has been more or less fixed. For the past few years, in Tokyo, we have been carrying out a 40-hour program for 11- or 12-year-old kids that allow them to understand their own community's urban environmental cultural resources. Like adults, kids do not have the complete knowledge of a town or city where they live. Through our program, children discover that their city is full of people, sounds, and landscape varieties as well as natural elements such as animals and not just buildings. In short, it lets participating school children go through the process of discovering urban environmental cultural resources themselves. This should massage their brains, which have been already fixed in terms of their thinking patterns, and make them aware of their own responsibilities in using their resources well.

4.2 Conclusion: China and Indonesia as Examples

We have, in various ways and forms, collected, listed, and presented as well as actually utilized urban environmental cultural resources of different Asian cities. In Japan, the researchers who belong to the preceding generation created a list of the architecture (Japan Institute of Architecture (1980)) built since the Meiji period (1868–1911). Based on their achievements, we have created a list of major examples of architecture built during the past 150 years in China (16 cities), Korea, Taiwan, Hong Kong, and Macau by collaborating with local researchers (Fujimori and Wan (1996)). Similar research programs were also carried out in Hanoi and other cities (Fujimori and Muramatsu (1997)). In China, the specialists there first responded to our surveys and their results, which led them to show their interests in buildings around them that they previously deemed insignificant. As a result, the Chinese society at large has started to recognize such buildings as their own heritage and utilized them as resources for tourism.

In Indonesia, similar surveys were carried out in Medan, Padan, and Jakarta. These surveys, which adopted the heritage butterfly model, are more sophisticated and more closely connected to the Indonesian society. In the year 2000, we established mAAN (modern Asian Architecture Network, http://www.m-aan.org), which allows us to link with each other across Asia in order to utilize urban environmental cultural resources of the whole region. While urban environmental cultural resources of a particular city belong to that city, the wisdom gained in each locality should turn out to be intellectual resources for all of Asia and beyond.

References

Fujimori T. and Wan T. (eds) (1996) *A Comprehensive Study of East Asian Architecture and urban Planning:1840–1945*. Taisei Kensetsu, Tokyo

Fujimori T. and Muramatsu S. (1997) *Preservation of Hanoi Architectural Heritage*. Construction Publishing House, Hanoi

Japan Institute of Architecture (1980) *Nihon kindai kenchiku sōron: nokoru Meiji Taisho Showa tatemono*. Gihodo Shuppan, Tokyo

Muramatsu S. (2007) "Kūkan bunka shigen no hyōka to sono keishō", in Fujino Y. and Noguchi T. (eds) *Urban stock no jizoku saisei*. Gihodo Shuppan, Tokyo, pp. 23–43

5. New Urban Strategy for Provincial Cities in Japan

Keisuke Fujii (Arranged and Translated by Mizuko Ugo)

5.1 Outline of This Lecture

Main issues:

(1) What is the historic stock? Definition.
(2) The Japanese Law for the Protection of Cultural Properties.
(3) What is meant by the reuse and regeneration of historic stock?

When intervening in an urban context, three main issues should always be tackled.

The first issue is what should be considered as historic urban stock and how it should be defined. In general, historic stock is thought to be tangible, a physical asset. But, this could not always be the case. Moreover, it is extremely important to understand what should be considered as the specific historic stock in the place we are working in. Historic stock is in fact always linked to the place to which it belongs.

The second issue relates to the concept of "cultural property" as defined in the current Japanese Law for the Protection of Cultural Properties. The purpose of this law, promulgated by the national government, is to preserve and manage the cultural heritage in a stable way.

Taking into consideration the current Protection Law, it is possible to underline two points. The first one is that the Law doesn't reflect the changes in the values attributed to the designated cultural heritage. In fact, in the concept of cultural heritage it is essential to link the object to a specific value. However, the value linked to the historic heritage might change, and sometimes even changes continuously. Despite this, the law establishes a definite, static, fixed link between the object and its value. Each cultural heritage has a definition unlikely to change in order to allow

Y. Fujino, T. Noguchi (eds.) *Stock Management for Sustainable Urban Regeneration*,
© 2009 to the complete printed work by Springer, except as noted. Individual authors
or their assignees retain rights to their respective contributions; reproduced by permission.

a much easier management of cultural heritage, including provision of financial allocations.

The second point is that unfortunately, the Japanese Protection Law is not linked enough with urban planning policies.

Finally, the third issue is how to reuse and regenerate the historic stock.

The above-mentioned three issues will be explained into two parts, using a case study of Joetsu City, Niigata Prefecture:

(1) The history of the Japanese legal system for the Protection of Cultural Properties.
(2) How the local administrations should select and deal with architectural properties?

5.1.1 The History of the Japanese Legal System for the Protection of Cultural Properties

1871 Proclamation by the Imperial Cabinet for the Protection of Antiquities
1897 The Ancient Shrines and Temples Preservation Law
1919 The Law for Historic Sites, Places of Scenic Beauty and Natural Monuments
1929 The National Treasures Preservation Law
1933 The Law Concerning the Preservation of Important Objects of Art, etc.
1950 The Law for the Protection of Cultural Properties

Introduction of new categories of cultural properties: Intangible cultural properties/Folk-cultural properties/Unexcavated archaeological cultural properties

The first law was issued just after what is called the Meiji Restoration, with the purpose of preserving the tangible cultural heritage. The next law, in 1897, is considered to be the starting point for the current preservation system; it stated that ancient shrines and temples and their treasures could be individually designated as national treasures. In 1919 a new law was promulgated to preserve historic sites, places of scenic beauty and natural monuments. In 1929 again another law was issued for the preservation of national treasures. While the 1897 law was only for ancient shrines and temples and their treasures, the 1929 preservation law enlarged its field of action including other buildings, such as privately owned houses, castles, and tea houses. After the Second World War, in 1950 the Law for the Protection of Cultural Properties was issued, on the basis of the previous laws. However, its scope was enlarged again to include new categories of cultural properties: intangible cultural properties, folk cultural properties and buried cultural properties.

5.1.2 Introduction of New Categories of Cultural Properties

1871	Tangible cultural properties: Objects
1897	Tangible cultural properties: Buildings and structures
1919	Historic sites (shell mounds, ancient tombs, sites of palaces, sites of forts and castles, monumental dwelling houses, etc.); Places of scenic beauty (cultural landscape); Natural monuments (natural sites and living species)
1950, 1954	Intangible cultural properties (artistry and skills); Tangible folk-cultural properties
1954	Unexcavated archaeological sites
1975	Preservation districts (historic cities, towns and villages); Traditional techniques for conservation of cultural properties

Each time that a new preservation law was issued, new categories of Cultural Properties have been introduced. Therefore, the objects considered to be heritage or cultural properties have increased constantly. Slide no. 3 shows those newly introduced categories.

After 1950, the revision of the law in 1975 introduced the new category of "Preservation Districts," i.e. historic cities, towns, and villages. If under the previous preservation laws it was possible to designate individual buildings only, for the first time the law allowed to designate historic areas as Cultural Properties, not only single buildings. However, the week point of this law is the lack of a strong link between the preservation of those historic areas and the urban planning process.

In 1996, a big revision of the law introduced the new concept of "Registered Buildings" (*Toroku bunkazai*), for buildings at least 50 years old. This system exists in many other countries outside Japan. Under this law, it is possible to register a building on a special list giving it recognized value under the law, though no financial support is given. Since this system started in 1996, around 40,000 buildings have been registered to date.

Again in 2005, the preservation law was revised and another category linked to urban planning and built environment was introduced, that of "Cultural Landscape."

5.1.3 The Categories of Cultural Properties
in the Current Protection Law

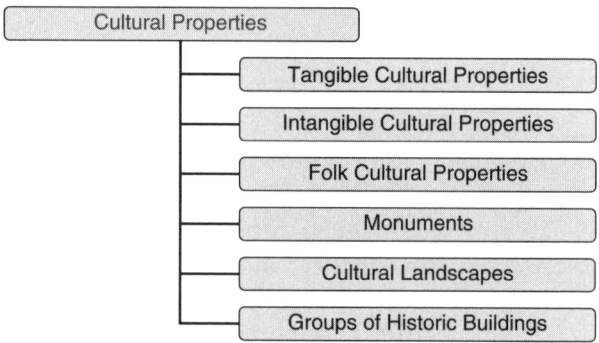

☆ **Conservation Techniques for Cultural Heritage**
☆ **Buried Cultural Properties**
- Buildings and other structures
- Historic sites, places of scenic beauty
- Rural landscapes
- Post towns, castle towns, farming and fishing villages

This plate shows the categories into which the concept of Cultural Property is divided. Among them, the categories of "Monuments," "Cultural Landscapes," and "Groups of Historic Buildings" are those strongly linked with urban stock, architectural heritage, and historic areas.

5.1.4 An Overview of the Selection and Designation
of Cultural Properties by Prefectures and Municipalities

Fig. 5-1 shows the number of buildings and other cultural properties designated at the prefectural and municipal level (as May 1, 2000). In particular, it is possible to notice that more than 2,000 "Buildings" have been designated as cultural properties at the prefectural level, while more than 8,000 at the municipal level. More than 2,000 properties have been designated as "Historic Sites" at the prefectural level and more than 12,000 at the municipal level. The "Preservation Districts for Groups of Historic Buildings" comprise 55 historic areas that are designated at the municipal level only.

Those listed cultural properties can be given financial support for their conservation and management.

(as of May 1,2000)

Type		Prefecture	Municipality
Tangible Cultural Properties	Buildings	2,318	8,312
	Fine and Applied Arts	8,837	37,364
Intangible Cultural Properties		157	1,024
Folk-Cultural Properties	Tangible	633	5,756
	Intangible	1,635	5,228
Monuments	Historic Sites	2,584	12,968
	Places of Scenic Beauty	234	997
	Natural Monuments	2,860	10,658
Preservation Districts for Groups of Historic Buildings		—	55
Cultural Properties Conservation Techniques		42	42

Fig. 5-1. Overview of the selection and designation of cultural properties in Japan

5.1.5 Outline of the Case Study: Joetsu City in Niigata Prefecture

When working in a specific place, it is crucial to think how to define and select the urban stock linked to that specific place, at the local level. Moreover, an adequate management system for this historic stock should be built up by the local administration.

Firstly, it is necessary to select criteria in order to properly sort out the urban stock. For instance, we could consider it as including all the constructions existing in the city. However, this is not possible in reality, because a value will always be attached to any of the existing buildings or areas and therefore a selection among the existing constructions will always be made. However, there are several types of values, which could be chosen at the national, local or even individual level.

Together with the value, it is also necessary to take into consideration the lifespan of the building. Depending on the building and its state of conservation, its lifespan could be still very long, or on the contrary quite short. At least 50 years since a building was built is the timeframe required by the Protection Law in order to evaluate the building as potential Cultural Property. However, in the reality there are cases in which a shorter period should be accepted because a great value could be attached to a building built much earlier than 50 years ago.

Of course, general definitions of Cultural Properties and evaluation criteria are necessary for the overall conservation system. However, at the same time specific situations at the local level should also be carefully examined through other criteria, which could help understanding and recognizing the value of each building at the local level, in that specific place.

The following case study, of Joetsu City in Niigata Prefecture, demonstrates that it is possible to apply specific criteria and management methods adapted to a specific place. This research started after a request made by the local authorities to prepare a preservation and management system for the local urban stock.

5.1.6 The American Bombing Project and Target Cities

Fig. 5-2 is of a map of Japan showing Joetsu City located in the northern part. This map shows the plan to bomb Japan prepared in 1944 by the US Army in Saipan and Iwo Island (Iwo Jima). The cities highlighted on the map are the targets of that project. It is well known that until the first half of the twentieth century, a huge urban stock of timber buildings built between the Meiji period and the early Showa period (1920s) still existed all over Japan.

Fig. 5-3 shows that the target of the American bombing were the most developed and populated towns, which are not those well-known nowadays for their historic districts. In other words, the towns that are not shown on this map were not bombed during the Second World War and today they are those where the historic districts still exist. This is the with such towns such as Takayama, Kurashiki, Matsue, Nakazawa, and also Joetsu.

5.1.7 The Survey and Preservation Project

The local administration asked our university laboratory to find a solution for two main issues: the empty plots of land in the city center and the *gangi*, arcades or corridors covered by a roof that are built all along the front of shops and dwellings. In fact, Joetsu is located in a heavy snowfall area and those covered paths made it possible to walk outside even in the case of heavy snowfalls.

However today, the City provides a public service to remove the snow and therefore those arcades are not anymore as useful as they used to be.

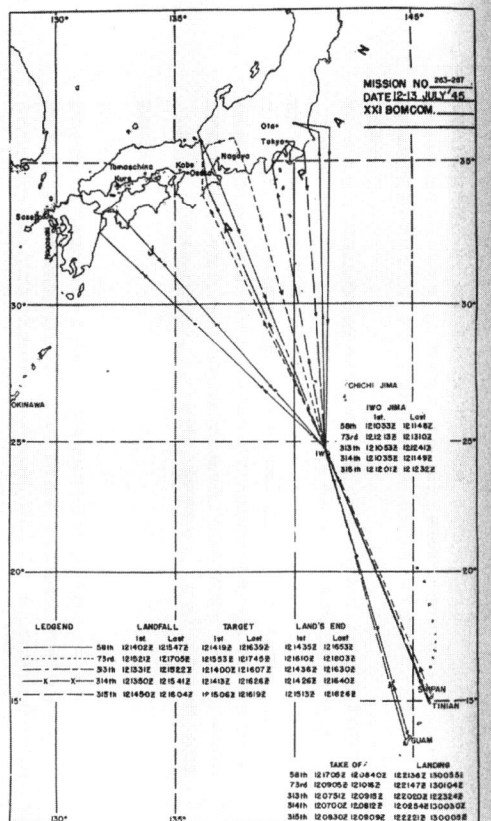

Fig. 5-2. US Air Force Route from Saipan to Japan

Nevertheless, there are no doubts about the historic importance of the gangi. which were once very common in northern Japan, but have been gradually demolished during construction work for new roads. Today, in Japan only two cities still preserve them: Kuroishi City in Aomori Prefecture and Joetsu City in Niigata Prefecture. Surprisingly, around 18 km of gangi are now left in Joetsu. Therefore, they have great value from the point of view of both their length (quantity) and history.

In fact, the gangi were built during the Edo Period when all the shops used to have those arcades in front of them. Therefore, it is possible to state that gangi represent the original and historic townscape of Joetsu City and in particular of its historic district: Takada District. The local government decided to preserve the gangi because they form the historic townscape peculiar to this district.

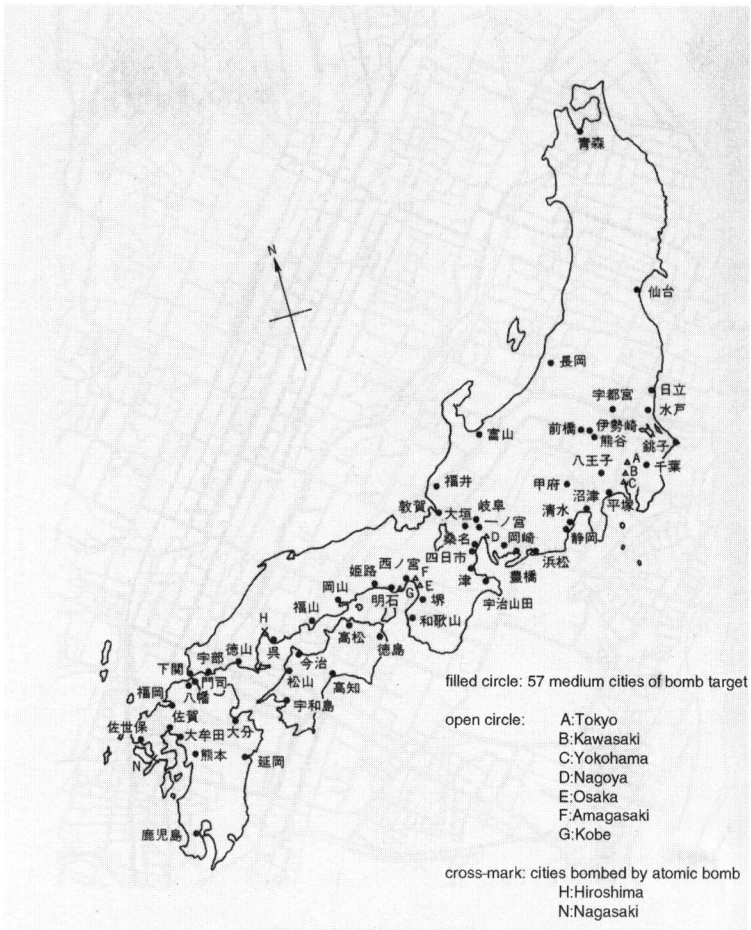

Fig. 5-3. Japanese cities bombed by US Air Force

Through the Protection Law it is possible to give value to an object or architecture with a view toward its preservation. However, the Protection Law recognizes value to ancient objects only. In the case of the regeneration project for Takada District, the issue to be faced was how to use this Protection Law in the best way.

In fact, the gangi are not that ancient because they have been continuously used and also repaired when damaged. Therefore, in order to preserve them the Protection Law was not sufficient and it was necessary to create a new preservation mechanism.

Moreover, gangi are useful devices that make it possible for pedestrians to walk in the streets while being protected from snowfalls. Therefore, it is not possible to preserve them in a static form, just as they are; any preservation should also allow their change and renewal whenever repair works would be necessary. It became clear that the Protection Law was not enough to protect the gangi, and it was indispensable to find a system that foresaw the possibility of changes.

It was thus necessary to go beyond the concept of Cultural Heritage as defined in the Protection Law and carefully take into consideration the specific values and characteristics of this heritage in order to find the best way to preserve it.

Fig. 5-4 is of a picture taken from the 1910s and 1920s, during the Taisho Period and early Showa Period. As can be seen, the gangi are tightly linked and pertain to the building behind them. If the building is pulled down the gangi also would be demolished. Therefore, to preserve the gangi it is vital to preserve the attached building as well.

Fig. 5-5 and Fig. 5-6 are pictures taken inside one of the houses. In the central part of the house, there is a typical open ceiling. This is considered to be the oldest house in the District. One can admire the refined interior design and impressive structure built around 100 years ago. It is said that in Takada Historic District around 5,000 houses built at least 50 years ago still exist.

Fig. 5-4. Gangi and townhouses in the city of Joetsu, *c*.1915

Fig. 5-5. Interior of townhouse in Joetsu

Fig. 5-6. Skylight of townhouse in Joetsu

In general, the houses in this area have a very simple exterior design and might even appear miserable looking because when damaged they are repaired using sheet iron to protect those deteriorated parts. However, on the contrary the interiors are skillfully designed and the structure is magnificent and strong.

The project started with the analysis of the actual situation. In dealing with the local administration, it was important to understand the objectives of the authorities concerned with implementing provisions related to both Cultural Properties and Urban Planning, and whether a cooperation system had already been established between the two of them. In fact, any conservation of architectural heritage should be carried out together with officials responsible for urban planning and urban regeneration.

It was possible to ascertain that the administration for Cultural Properties had a strong will to preserve the gangi through the establishment of a specific policy. Moreover, a field study was carried out by the Laboratory's students, with six volunteer citizen-surveyors participating. The method and guidelines already defined for the analysis and survey of shop buildings were followed. The citizen-surveyors investigated the urban environment and the characteristics of the townscape and the architectural details, collecting data on what they found interesting. It is possible to say that all the information collected during the field work is part of the urban stock. Therefore, it can be understood that if more and more people participate in such surveys, the categories of the stock can vary and increase in typologies and forms.

5.1.8 Proposed Reuse and Conversion Projects

The field analysis offered the opportunity to ponder once again what should be considered as the historic stock of the city. In order to preserve it, the study group suggested practical solutions especially for the reuse and/or conversion of town-houses and empty plots of land.

Fig. 5-7 to Fig. 5-10 show examples prepared by the students of proposed changes in the use of existing or of newly designed buildings. Fig. 5-7 shows the plan for converting three townhouses into a nursing home. Fig. 5-8 shows the plan for reusing the historic timber structure of a townhouse's central open ceiling while redesigning completely the interiors. Fig. 5-9 shows the plan for a new construction to be built on an area made from several empty plots of land within the historic city. The functions proposed for this new building include public services and a clinic. Fig. 5-10 shows the proposal for linking together many townhouses to be transformed into a shopping mall.

Fig. 5-7.

Fig. 5-8.

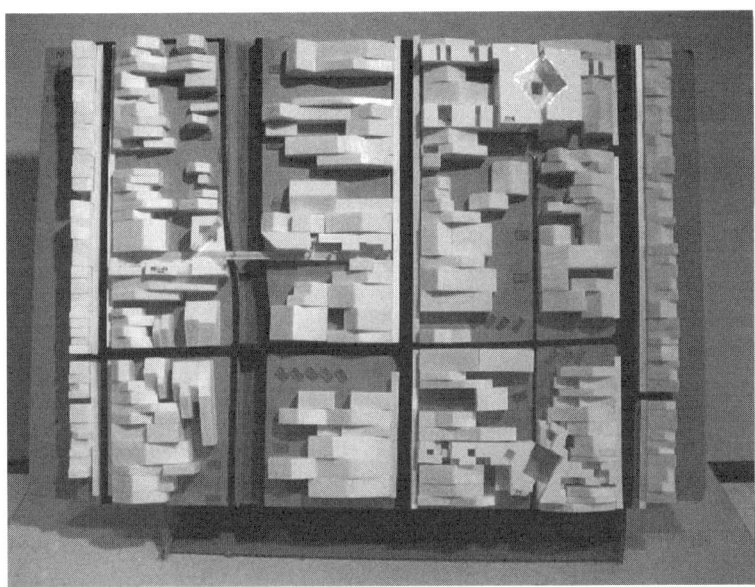

Fig. 5-9.

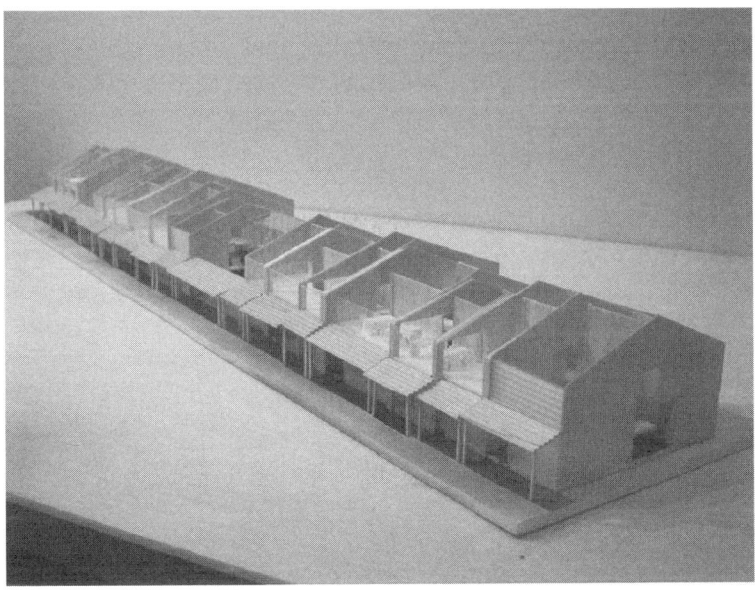

Fig. 5-10.

All the proposed projects tried to preserve the historic buildings while suggesting new and useful functions for them. The solutions proposed have been a response to the question of how to use the Protection Law with the purpose of developing built areas or specific districts through community action (i.e. bottom-up action), while preserving the existing historic buildings.

5.1.9 The Timeframe for the Regeneration Project

A timeframe is here suggested in order to carry out the regeneration project.

Taking into consideration that the implementation of the project takes a long time and that public financial support cannot be kept unspent if committed to a project for too long, a spending plan has been carefully formulated.

The timeframe for the implementation of the project consists of three parts: short, midterm, and long term. The overall project has been divided into six components to maximize effectiveness. This Takada District Project is now ongoing; at the moment, one of the townhouses is being converted into a community center.

5.2 Conclusion

Though Japan's urban planning is typically thought of as a "pull down and rebuild"/"demolition and reconstruction" process, this is not always the case.

In this situation, instead of pulling down the existing urban stock to replace it with new buildings, the existing architecture with an already very long life is now being reused. Moreover, this regeneration of the urban stock also becomes a way to preserve and manage the townscape and the urban environment as a whole. Finally, it is possible to state that the conservation of the city's heritage is a valuable resource in the regeneration of the city.

Architectural heritage should thus be used as a positive driving force in a city's regeneration. Of course, changes and urban transformation should always be allowed. However, those changes should be regulated and well-managed, so they can happen more slowly. The next step will be trying to resolve the gaps existing between the Protection Law and the issues involved in actual implementation.

6. Life-Span Simulation of Reinforced Concrete Structures: Toward Rational Stock Management

Tetsuya Ishida

6.1 Introduction

Life-span simulation of reinforced concrete structures for rational stock management is a hot topic of today's civil engineering research and development. This technology is now being applied to actual structures for predicting residual lifetime of existing structures or the rational lifetime for newly constructed structures.

Concrete is a composite material consisting of cement, sand, and gravel, and is most the widely used construction material around the world. The main constituents of concrete are calcium and silica. They exist all over the earth, unlike other materials like rare metals, which are not abundantly available in all regions. This ease of availability and production is a big advantage of concrete, but on the other hand, the same may lead to disadvantages due to the failure to observe quality control. In order to make very good quality concrete structures at a cheap price, not only good development methods, but high-quality construction, maintenance, and management systems are necessary. For this reason, accurate predictions of concrete quality and its durable service life are necessary.

6.2 Developmental Phases in Civil Infrastructure

Much infrastructure in Japan was constructed during the high-growth period of the 1960s and 1970s. About one fourth of all bridges were constructed during these decades. Therefore the average age of the bridges in Japan is now 37 years old, but 20 years from now, the average age would be 57 years if no new infrastructure were constructed. Therefore a lot of rehabilitation, retrofitting, replacement, and maintenance are expected, and those costs will increase in the future. In addition, rapid construction may lead to poor quality, hence the need to develop some technology for rational stock management.

Figure 6-1 shows the per capita consumption data of cement from 1950 to 2000 (Ouchi 2007, and Figure 6-2 shows data about the age of bridges in Japan. The maximum amount of cement consumed in Japan during this period was in 1972–1973, when it was more than 700 kg per capita. After that the consumption rate does not change so much. It is not a direct index of the amount of infrastructure built, though a rough estimation can be made from cement consumption. Singapore follows the same trend as Japan, whereas Taiwan lags 10 years behind. Korea is more than five years behind; Malaysia follows, with almost the same trend, and Thailand is 20 years behind Japan. In the case of China, since its economic rate is high, construction of infrastructure is therefore in abundance.

From these data, it is evident that since concrete consumption is huge, the impact of even little improvement in concrete's material technology is great. Even a little cost reduction is thus very much significant. Japanese society is now aging and there is a lower birth rate, so there is a need to build good quality infrastructure with minimum cost. No negative inheritance should be passed to the next generation, and therefore not a lot of money can be spent for maintenance and repair of infrastructure. This situation may occur in other Asian countries in the future, as development progresses with cement consumption.

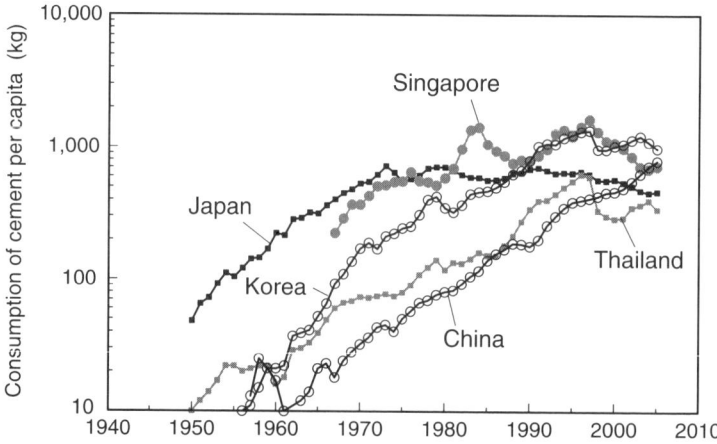

Fig. 6-1. Cement consumption rate in Japan

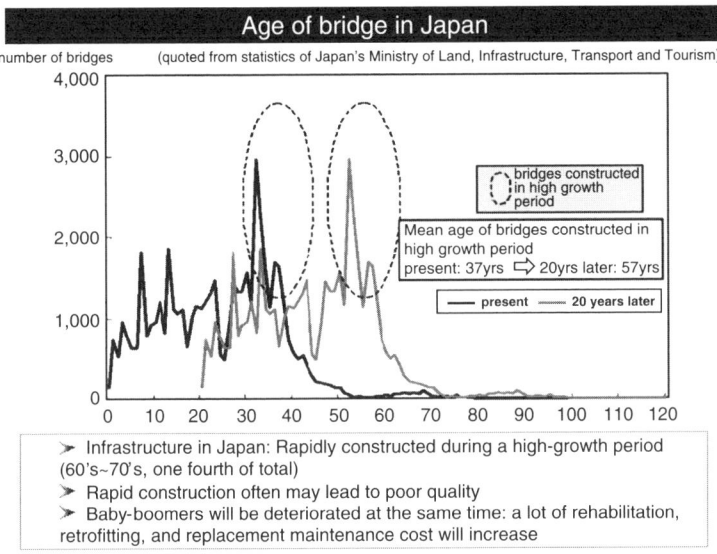

Fig. 6-2. Bridge age in Japan

6.3 Life-Span Simulation for Materials and Structures

From this social background, the need at present is to develop the life-span simulator of concrete structures. Cement and concrete materials include lots of pores from nano to millimeter scale, and the existence of microscopic pores affects the macroscopic properties of concrete. Recently, numerical simulators based on physical properties have been developed by a couple of research groups in Japan, the United States and the Netherlands since the early 1990s. This technology is being used for engineering applications, with venture companies founded in Japan, the US, and France.

Figure 6-3 shows the schematic representation of lifespan-simulation systems developed by the Concrete Laboratory at the University of Tokyo. Two numerical tools have since been developed. One is the thermo-hygro system, which can deal with cement hydration, microspore structure development, moisture transport and chloride transport under given environmental actions. The other is the structure-analytical system, which predicts the structure behavior under static and seismic loads. The performance of concrete will change after casting and during its service lifetime, so by using these numerical systems, the deterioration starting time can be predicted via computer. The two systems are interlinked in order to deal with structure and material properties (Maekawa et al. (1999)).

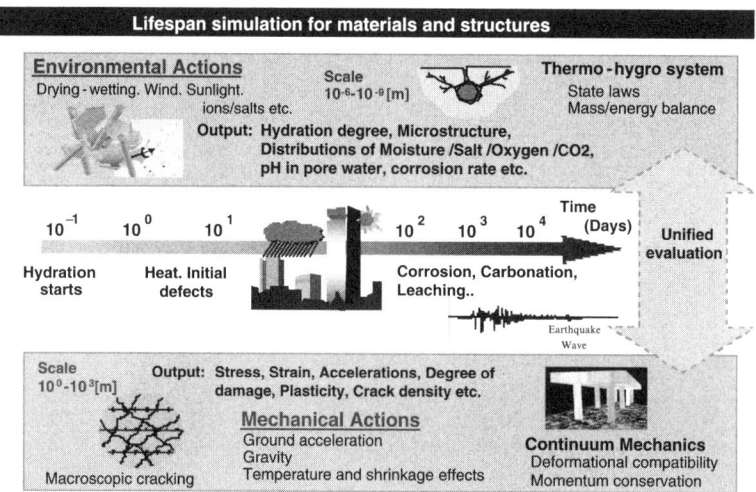

Fig. 6-3. Schematic representation of simulation systems for materials and structures

6.4 DuCOM: A Simulator Program for Service Life Prediction

With DuCOM, mass and energy balance equations are solved, which are the governing equations in Finite Element (FE) analysis. The equations include potential, flux and sinks terms, and are solved in terms of temperature, pore-pressure (moisture content in cement materials), chloride concentration, carbon dioxide, oxygen, calcium, and chloride, whose relationships can be used to predict the long-term deterioration phenomenon. In this system, the size of target structure, shape, mix proportions and initial and boundary conditions are provided. Mass and energy balance equations are then solved for cement hydration, and from this model the temperature and hydration level of each component are predicted (Ishida and Maekawa (1999)); (Ishida et al. (2007)); (Maekawa et al. 2003).

Figure 6-4 shows the flow chart representing the basic framework of DuCOM and its sub-models. The development of multiscale micro-pore structures at an early age is obtained for the average degree of cement hydration in the mixture. For any arbitrary initial and boundary conditions, the pore pressure, relative humidity, and moisture distribution are mathematically simulated according to a moisture-transport model that considers both vapor and liquid phases of mass transport (Ishida et al. (2007)). The moisture distribution, relative humidity, and micro-pore structure characteristics

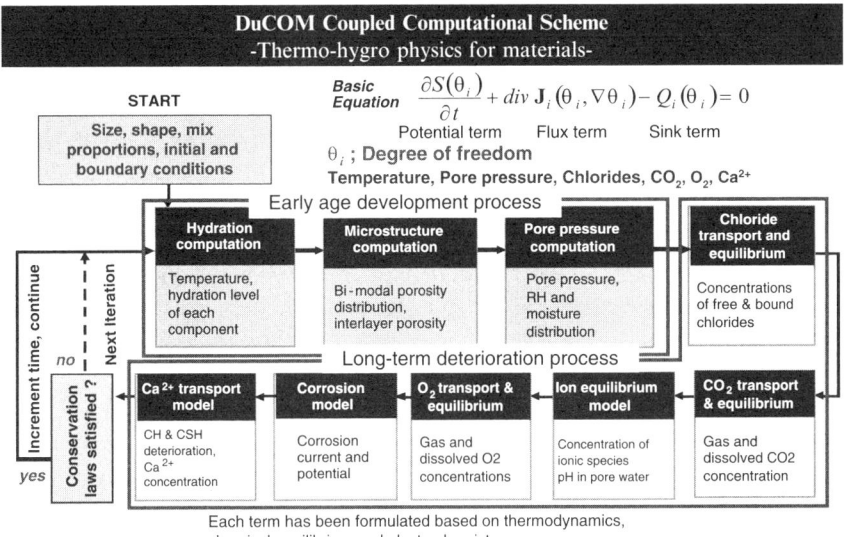

Fig. 6-4. Basic framework of DuCOM

in turn control the chloride, carbon dioxide and oxygen diffusion and rate of carbonation under arbitrary environmental conditions.

From this model a prediction is made for dense or coarse micro-pore structure, and this information is made available to the moisture-computation model. If the micro-pore structure is of very poor quality the moisture is easily transported, so this information is very important for moisture-transport computation; from such data the relative humidity in the pore system is computed. Actually there are two types of micro-pore structure: one involving capillary pores whose radius is one micrometer, and the other a gel pore with a very fine radius on the nanometer scale. This information is used to simulate capillary, gel, and interlayer porosity, and the moisture equilibrium inside such nano scale pores is computed. These models as described above correspond to the early age development process.

The corrosion of reinforcement in the structures is one of the biggest problems faced by engineers. since this causes the durability of the structure to deteriorate. The corrosion is caused when the chlorides in the system break the passive layer on steel. Therefore to predict corrosion, chloride transport and an equilibrium model is required (Hussain and Ishida (2007)); Iqbal and Ishida (2007a, b)). Carbonation is another factor that causes corrosion and hence the need for the carbon-dioxide transport model and carbonation model (Maekawa et al. (2003)). Overall corrosion rate is effect by oxygen supply, so the oxygen-transport model has been included in the

DuCOM system. Finally, to insure very long-term durability for nuclear waste structures, the calcium-ion transport model has been installed in this numerical system (Nakarai et al. 2006).

This prediction is very important to assess the potential durability of concrete for more than 10,000 years. In such long terms, the prediction of the destruction of micro-pore structure CSH (Calcium Silicate Hydrate) gel is very important because the calcium ions dissolve in pore water that comes from CSH, and this leads to the deterioration of CSH itself.

By the input of material basic data such as cement content and constituent's percentages, water content, aggregate content, air content, and boundary conditions such as humidity, temperature, and curing conditions in the DuCOM numerical system, the simulation of the material and structural performance over time and space is carried out.

There are a lot of equations in the DuCOM system, but mass balance and energy balance equations are the governing ones, and each term has been formulated based on each physical phenomenon. A DuCOM Windows version for 1D durability analysis has now been released, with a user-friendly Graphical User Interface (GUI) as shown in Figure 6-5. It can simulate one-dimensional mass and energy transport mechanisms. Practitioners are using this system for durability design of concrete structures. Figure 6-5 shows a screen shot of the Windows based version of DuCOM.

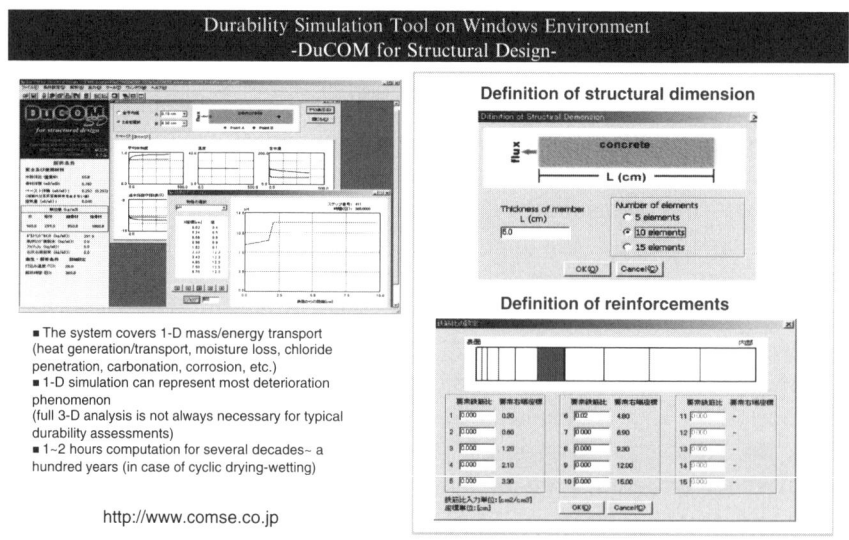

Fig. 6-5. Screen shot of Windows-based DuCOM

6.5 Modeling and Verification at Early Age Development

The verification was done even at different casting temperatures as shown in Figure 6-6. Temperature is very important while accessing the cracking due to cement heat. At casting temperatures of 10 °C, 20 °C and 30 °C with and without fly ash as admixture, it was observed that the proposed model could nicely predict the real data. So this kind of information is used to predict early age cracking due to thermal stress (Maekawa et al. 1999).

For the prediction of moisture loss at early age under severe conditions, an experiment was conducted in which the mortar specimens were put in a vacuum chamber, and then a measurement for mass loss was taken periodically as shown in Figure 6-7. The moisture loss due to drying with time was determined analytically and compared with experimental results. The effect of water cement ratio for 0.25 and 0.32 was also checked, and it was found that for samples with high w/c, the moisture loss is more.

The prediction of moisture loss behavior under high and low temperatures was experimentally verified by subjecting prism specimens 4" × 4" × 16" in size to 20 °C and 60 °C temperatures at water contents of 50 and 25 percent as shown in Figure 6-8.

This very fundamental experiment must be combined with a performance evaluation of concrete structures. In the preceding sections a brief introduction has been provided to predict an early age development process. Hydration and micro pore-structure and moisture models are used to simulate the temperature rise, moisture loss behavior and micro-pore

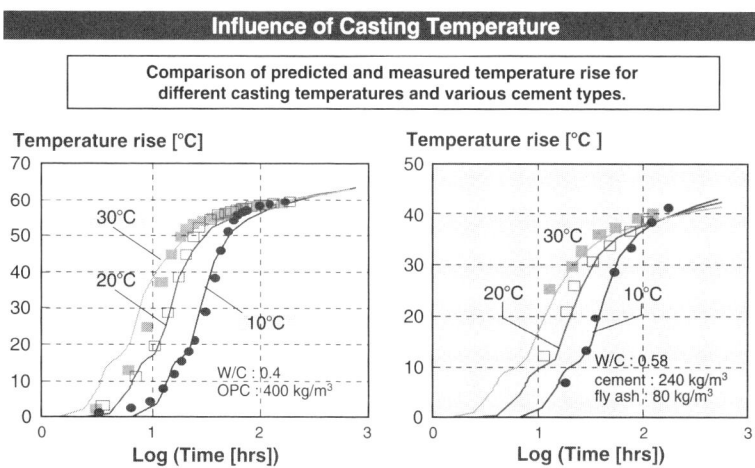

Fig. 6-6. Influence of casting temperature on temperature rise in concrete

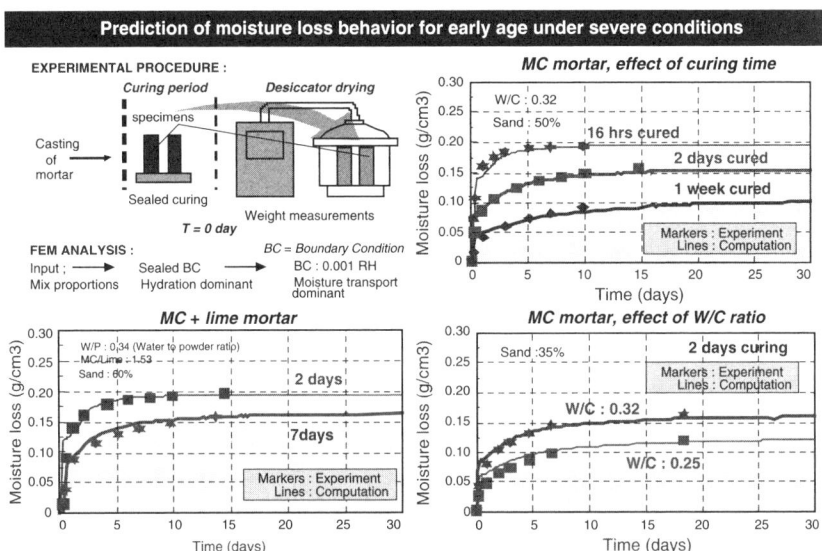

Fig. 6-7. Moisture loss at early age under severe environment conditions

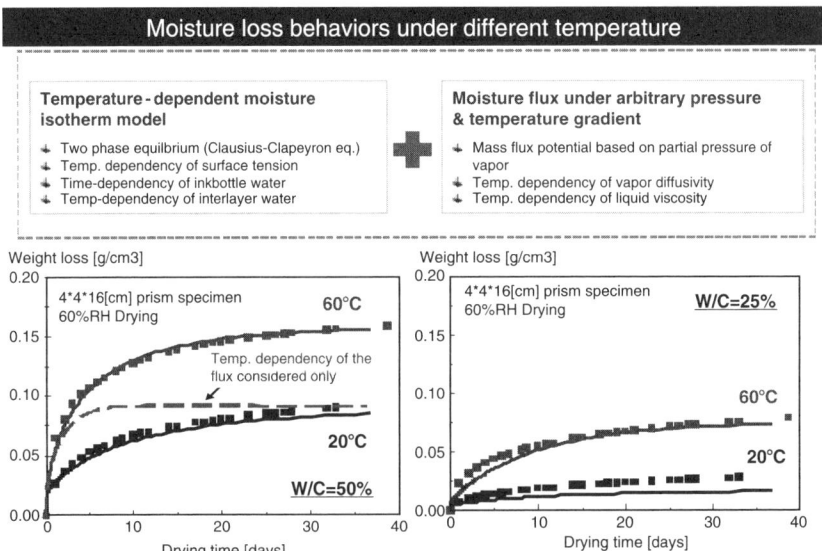

Fig. 6-8. Moisture loss under various temperatures

structure development, which is very important for predicting initial defects such as thermal cracking and shrinkage- induced cracking. To evaluate the initial defects by this computation model, an early age development model is needed. Long-term deterioration phenomena such as steel corrosion carbonation and calcium leaching are also closely linked with early age development processes.

6.6 Modeling of Creep and Shrinkage

In this section a brief strategy is introduced to simulate the creep and shrinkage in the DuCOM system as shown in Figure 6-9. Since shrinkage and creep are strongly affected by the water content, an amount of liquid and vapor moisture in the system is therefore required. Also at an early age, stiffness is dependent on hydration levels. The micro-level information is applied to this multi-scale constitutive model, which offers instantaneous, visco-elastic, visco-plastic and plastic deformations.

Verification has been made for the prediction of autogenous and drying shrinkage. When the specimen is exposed to drying, it starts to shrink and the amount of shrinkage is dependent on mix proportions and drying conditions. Specimens were tested with various water contents, and the experimental results closely match with the analytical results. Experimental verification of basic and drying creep was also carried out. And now this model has been applied to the real structures.

Last year very severe damage was found in an actual PRC bridge with seven spans and box girder cross-section, as shown in Fig. 6-10. It was located in Wakayama prefecture in the middle part of Japan. The construction was started in January 2001 and completed by April 2002. After one year, by October 2003, numerous cracks and large deflections were found in this bridge. As a consequence, JSCE set up an emergency committee to evaluate its structure performance and to learn the reason for such huge cracking. DuCOM was used to find the source of cracking in that bridge. The analytical input (water content, cement content, type, aggregate content, environmental conditions like temperature change in winter and summer, and reinforcement ratio (was provided as per designed specifications.

From this analysis, it was observed that after 700 days there was a sudden reduction in shrinkage strain due to member response, which shows the generation of cracking. The analytical results shown in Figure 6-11 roughly correspond to the actuality. Through this analysis it was found that the high amount of steel in RC members, plus significant shrinkage caused by the

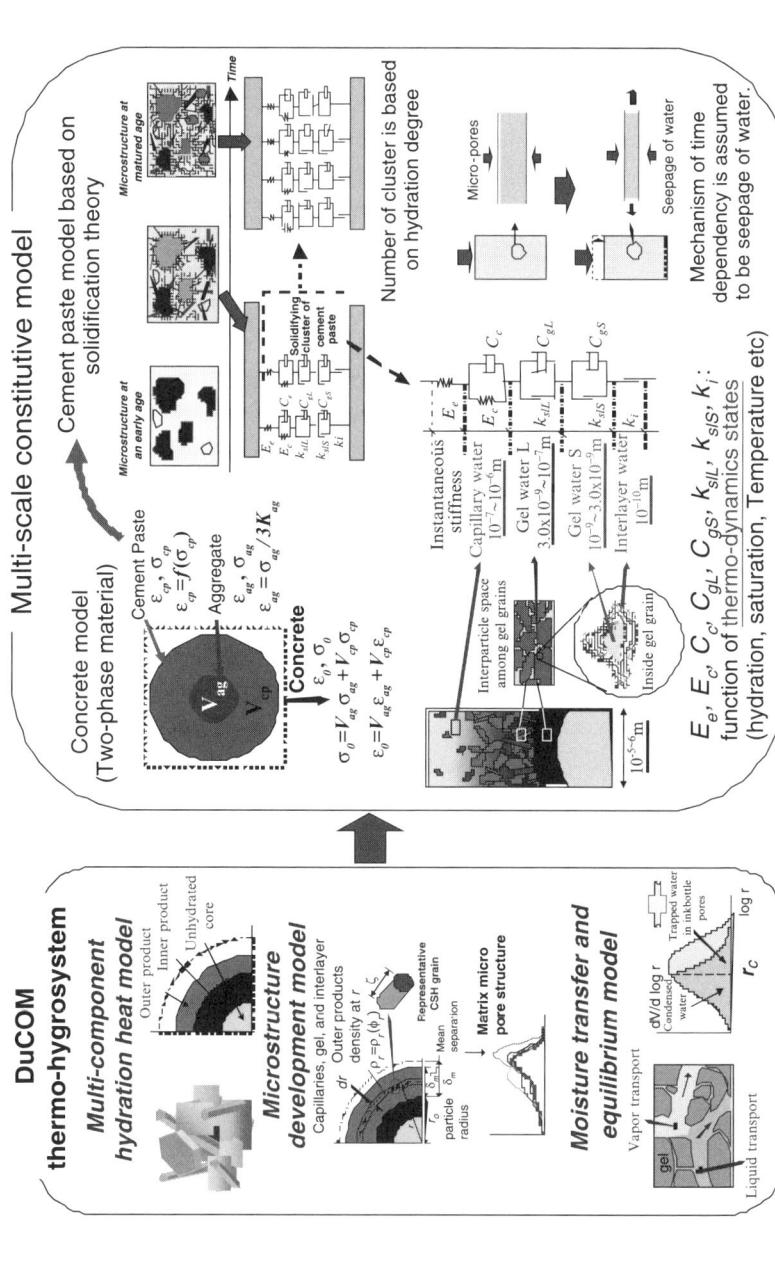

Fig. 6-9. Multi-scale constitutive model for creep and shrinkage

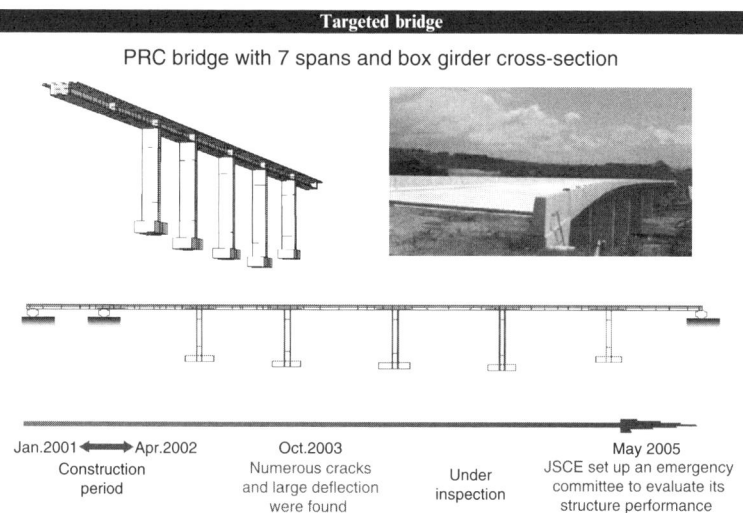

Fig. 6-10. Diagrammatic sketch of PRC Bridge

specific mix proportions and types of aggregates, caused numerous cracks and large deflections.

6.7 Modeling of Chloride Transport

In this section, chloride transport is discussed as factors involved with the maintenance and service lifetime prediction of concrete structures. Simulation of steel corrosion in concrete exposed to marine environment was done.

In this analysis the target structure was a T-shaped RC girder bridge that was exposed to marine environmental conditions as follows: CO_2 concentration of 0.07 percent, O_2 concentration of 20 percent, and Cl^{-1} concentration of 0.51 mol/L, with cyclic wetting and drying conditions as shown in Figure. 6-12.

Because the bridge is in a marine environment, the structure is sometimes exposed to cyclical drying and wetting, which was taken into consideration in this analysis. Over time, the chloride ions can penetrate inside the RC member and gradually corrode the reinforcement. Three types of analytical computations were performed (at W/C 40%, 50% and 60%) with 60 mm cover depth. From this analysis it was predicted that for 40 percent water content, the corrosion-induced cracks will start to appear after 52 years, thereby indicating the right time for rehabilitation. Similarity for 50

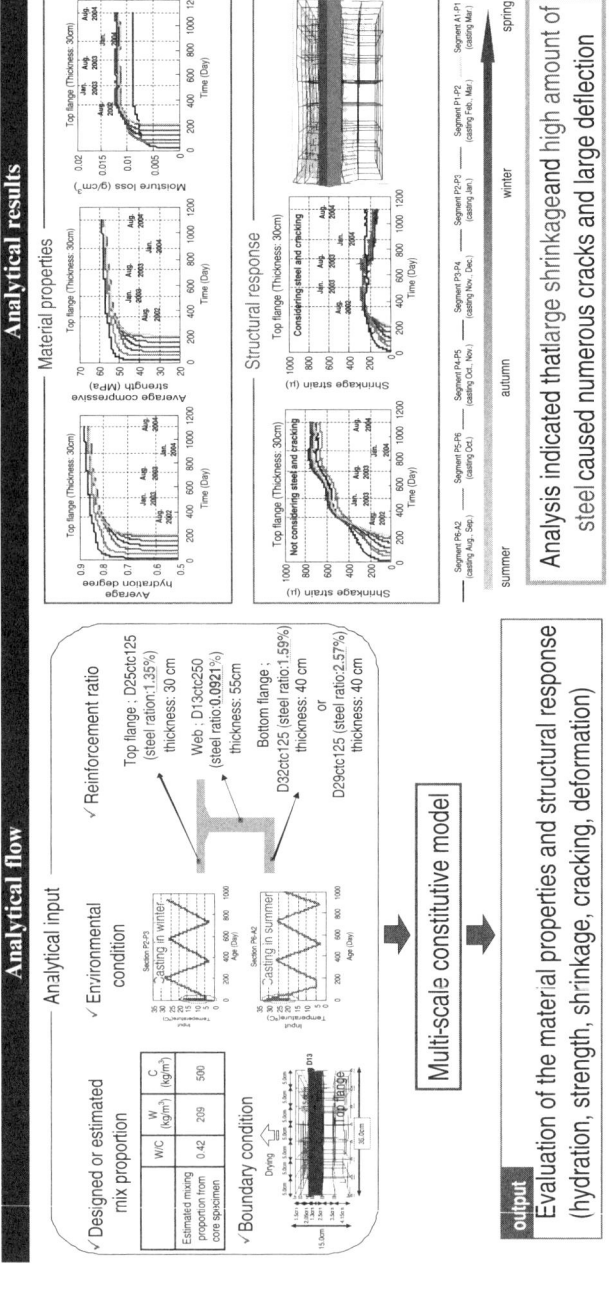

Fig. 6-11. Analytical results of PRC Bridge

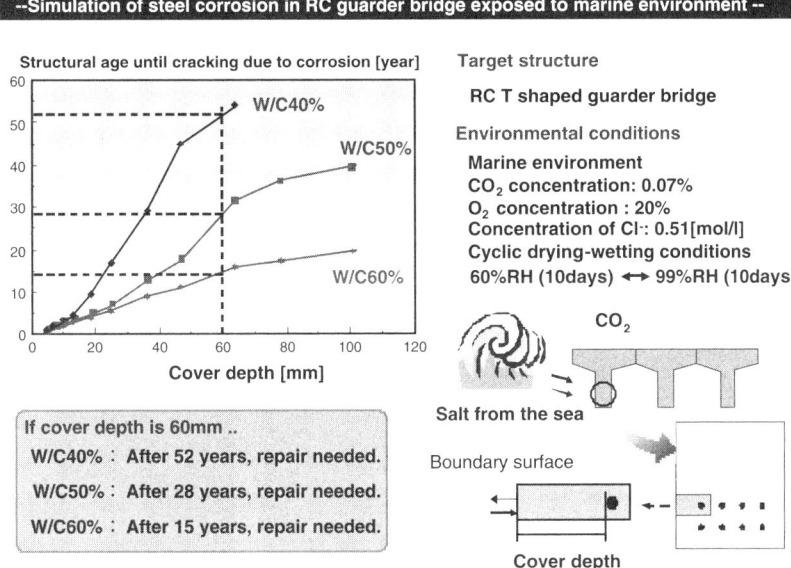

Fig. 6-12. Performance evaluation of RC structures

and 60 percent water content, repairs would be needed after 28 and 15 years respectively.

When the concrete has a high water-to-cement ratio, the diffusivity of chloride and oxygen is high, which means less resistance against corrosion, and early repair is required. If this kind of behavior is known at the initial construction stage, it will lead to a very rational design as it includes the durability aspects with structure performance and serviceability.

The governing equations in DuCOM for chloride transport has three terms as shown in Figure 6-13: potential, flux and sink. Potential term means how much chloride can exist in a unit volume of the cementitious material, and for this porosity and saturation is required. Flux term refers to the movement of chloride ions in cementitious material due to diffusion driven by ionic concentration difference, and convection due to transport of chloride by moisture movement (Iqbal and Ishida (2007a, b)); (Maekawa et al. (2003)).

Since the target is a very complicated micro-pore structure, some parameters such as constrictivity and tortuosity are therefore included, which represent the complexity of micro-pore structures. Tortuosity is the reduction factor in terms of complex micro-pore structure, and constrictivity takes the effect of pore radius and density of ions in the geometrical shape

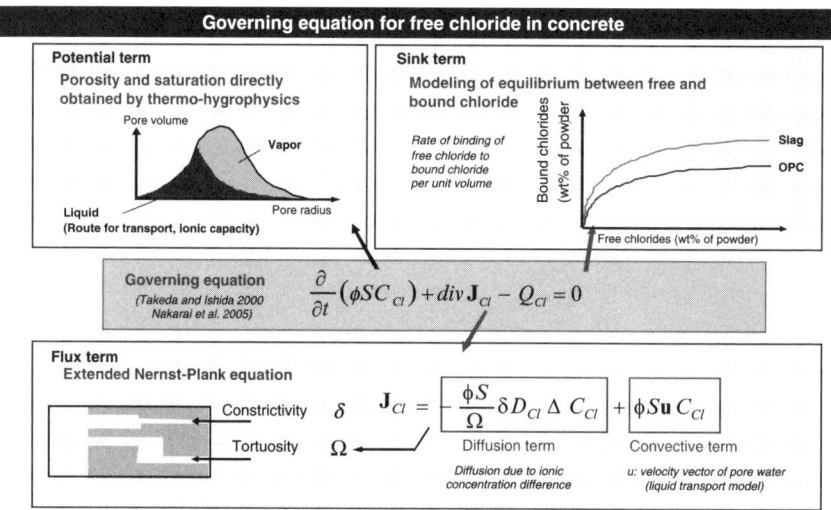

Fig. 6-13. Governing equations in DuCOM for chloride modeling

of micro-pore structure. Sink term, which describes the rate of binding of free and bound chlorides, uses a model for equilibrium between free and bound chloride.

The chloride transport model was verified by experiment for different types of mortars containing fly ash and blast furnace slag. The specimens were submerged under salty water, and after desired exposure they were picked up and tested for chloride content. The experiment and analysis indicated a very nice agreement, but that experiment was conducted under constantly submerged conditions.

To verify the model for alternate wetting and drying environments, experimentation on chloride behavior under various cyclic wetting and drying conditions were carried out under laboratory-controlled conditions (Iqbal and Ishida (2007b)). In order to produce such a complex environment, the specimens were exposed to a weekly hygral cycle comprising $x$hr wetting in three days, and $x$hr wetting in four days, where x is the time of wetting period in hours. For this experiment three wetting exposures (with $x = 1$, 9, and 33 hours) and a completely submerged case for benchmarking were considered (Iqbal and Ishida (2007a, b)).

Figure 6-14 shows the experimental results of the verification of the chloride model for alternate wetting and drying environments. From this experiment, it was observed that the results for the submerged case are in good agreement with the analytical results. In the 33 hours wetting case, a peak was observed due to two-way chloride movements during drying, and

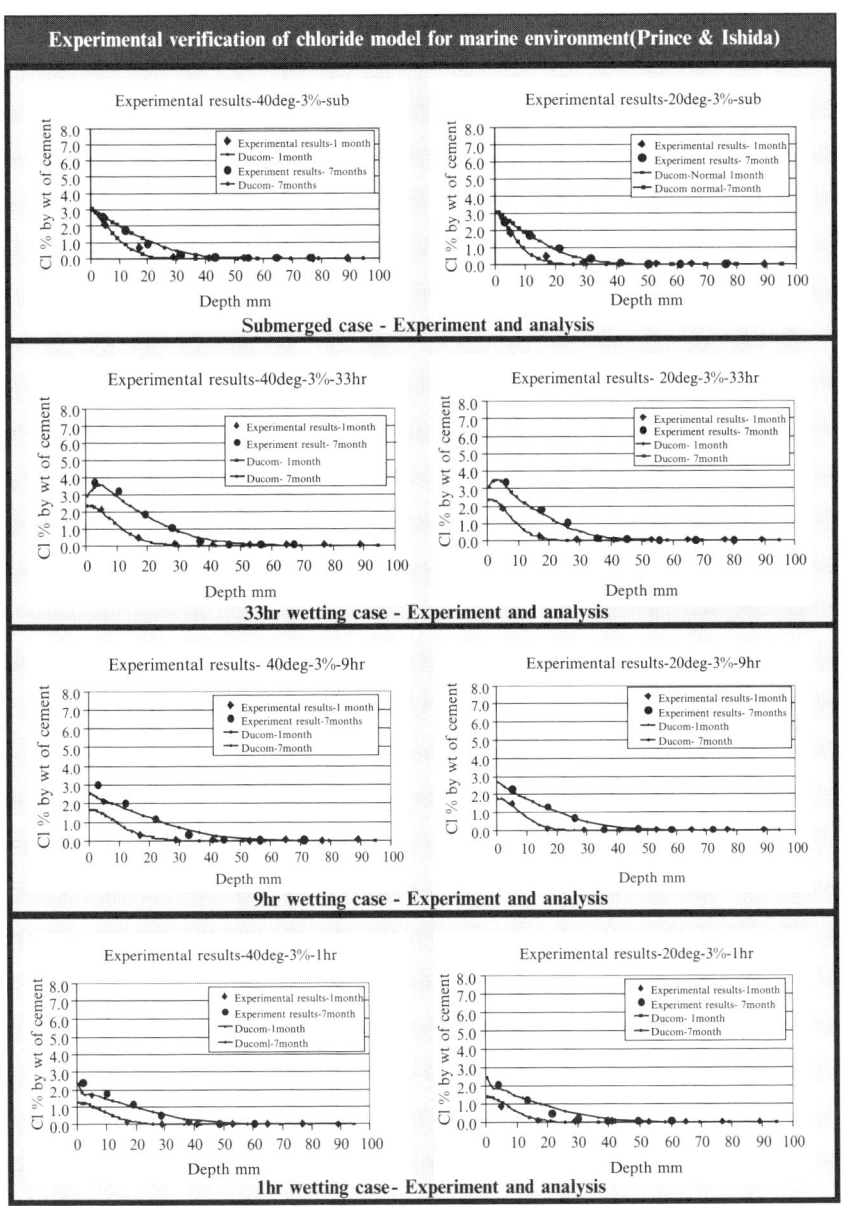

Fig. 6-14. Verification of chloride model for alternate wetting and drying environment

results were nicely predicted by the model analysis. The same can be said for the nine hour and one hour wetting cases. Under well-controlled laboratory conditions the model can predict very well. Hence there is a need is to check its applicability on real structures in a real environment.

6.8 Performance Evaluation of Existing Structures

When the target makes use of very well-controlled experimental data, the model predicts very well. For example if the mix proportions and environmental conditions are known, the system can simulate the chloride profile, carbonation, moisture-loss behavior, and so on. But when the target is a real structure, the situation becomes very difficult. When the newly constructed structures or small specimens in the laboratory are simulated, it is very easy to know the requisite abovementioned properties. But the real problem is the identification of these properties in existing structures.

Figure 6-15 is a schematic sketch of performance evaluation of new and existing structures. In the case of performance evaluation of a new structure, the cement content, admixture, and water content are known, and the simulation can start from initial construction; but when the target is an existing structure, the simulation has to be started from some point in the timeframe,

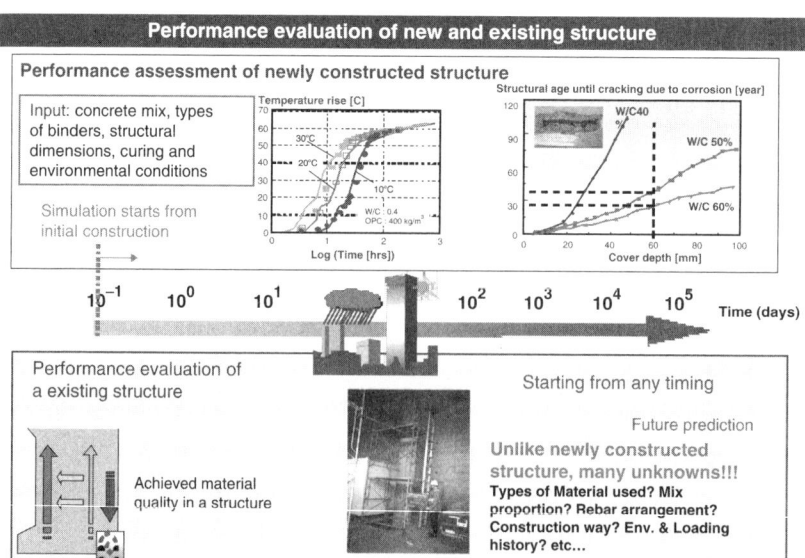

Fig. 6-15. Performance evaluation of structures

like 10, 20, or 30 years being already in service. Unlike the case with newly constructed structures, there are many unknowns here. Sometimes the design documents, which include mix proportions, dimensions and construction method, may be helpful in estimating those initial conditions.

Besides this, environmental conditions from the time of casting and during service life must be considered. These include such factors as temperature, surrounding water, humidity, and water leakage, oxygen, carbon dioxide and chloride concentrations. In some cases the weather database at each specific point may be obtained to infer the pattern of temperature, moisture, and chloride history for that specific structure, but in many cases it is very difficult to estimate this kind of environmental condition. Not only this, sometimes design documents are also not available.

To solve this problem, the unification of non-destructive tests (NDT) and numerical analysis is being considered, since NDT technology is much enhanced. The permeability or diffusivity of cover concrete and mass transport characteristics can be obtained by NDT to identify the input for simulations. The other way is by taking a core sampling of real structures and then estimating the initial mix proportion and material quality. This information about material properties, environmental conditions, and load history is used as input for DuCOM (material simulation system) and COM3 (structure simulation system) and the present state can be reproduced, and extended for future prediction. A possible crosscheck can be made by comparing the simulation of the present state with the real data obtained from the field. According to this methodology, the carbonation of the railway in Tokyo was simulated (Ishida and Li et al. 2008). Figure 6-16 shows the experimental verification results for the carbonation model. The laboratory level verification of this model has been done under different mix proportions and different temperature conditions.

6.9 Unification of NDT and Simulation Technology for Existing Structures

Focused research is now being implemented to develop a unified system to simulate the strength and durability of existing infrastructure. To achieve this objective, nondestructive testing (NDT) techniques and simulation technology shall be utilized in the future.

As an example, consider the Torrent method, which consists of a vacuum pump and two chamber cells to be placed on the concrete surface for testing. After the two chamber cells are attached on the concrete surface, the pressure

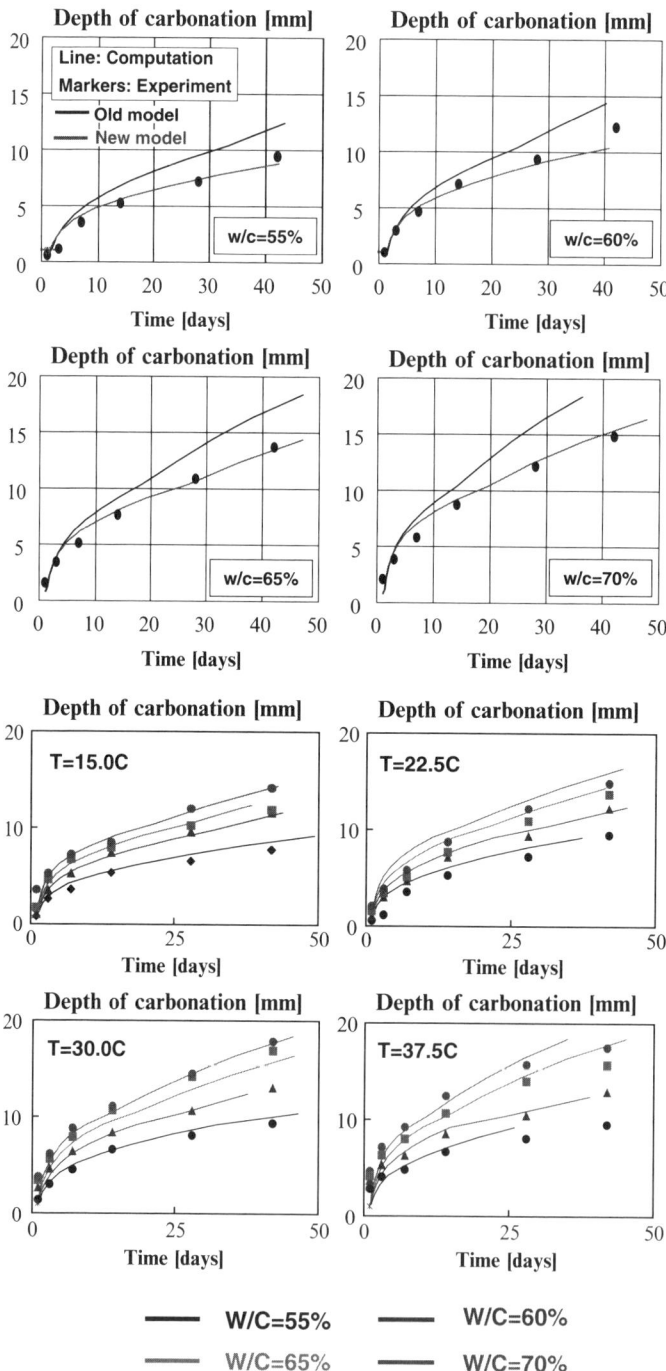

Fig. 6-16. Verification of carbonation model

inside the cell is reduced by the vacuum pump. Due to this reduced pressure the air flows from the concrete to this chamber (Torrent (1992)). A detailed sketch of Torrent's method is shown in the Fig. 6-17.

If the concrete has very dense micro-pore structure then it has low diffusivity and this airflow is much less, resulting in a slow increase of this air pressure inside the cell. However the change of pressure would be high for poor-quality concrete due to large diffusivity, resulting in much airflow from the concrete to the vacuum chamber. Measuring and correlating the airflow into the chamber measures the diffusivity of cover concrete. Since this apparatus is just attached on the surface, it is thus not necessary to destroy the concrete. In a similar way other NDT tests can be used to determine strength, moisture content, etc. and these values may then be used as input in the simulation system.

6.11 Conclusion

The estimation of material properties and environmental actions throughout the service lifespans of both the existing and newly constructed structures is very important for simulation testing that can contribute to stock management. In the future, focused research needs to be done to develop a rational approach for modeling of environmental actions in a manner that works in tandem with the material- and structure-modeling systems. The envisioned unification system that would integrate NDT and DuCOM is a direct use of the values obtained from NDT as input for DuCOM for determining durability aspects of existing structures.

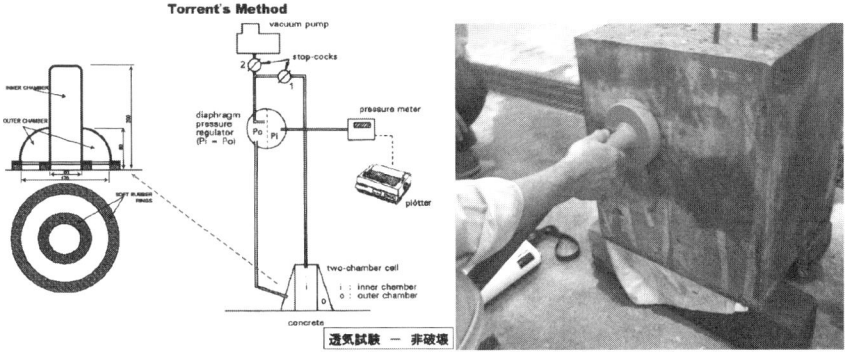

Fig. 6-17. Torrent apparatus

References

Hussain, R.R. and Ishida, T. (2007). Modeling of corrosion in RC structures under variable chloride environment based on thermodynamic electro-chemical approach. The International Symposium on Social Management Systems, Yichan, China

Iqbal, P.O. and Ishida, T. (2007a). "Chloride transport in concrete exposed to marine environment", *JCI Annual Convention*, 29(1)

Iqbal, P.O. and Ishida, T. (2007b). Modeling of chloride transport in concrete coupled with moisture migration in marine environment based on thermodynamic approach. The International Symposium on Social Management Systems, Yichan, China

Ishida, T. and Maekawa, K. (1999). "An integrated computational system for mass/energy generation transport and mechanics of materials and structures", *Journal of JSCE*, 627(44)

Ishida, T. et al. (2007). "Enhanced modeling of moisture equilibrium and transport in cementitious material under arbitrary temperature and relative humidity history", *Cement and Concrete Research*, 37:565–578

Ishida, T. and Li, C. (2008). Modeling of carbonation based on thermo-hygro physics with strong coupling of mass transport and equilibrium in micro-pore structure of concrete, Journal of Advanced Concrete Technology, 6(2):303–316.

Maekawa, K., Chaube, R. and Kishi, T. (1999). *Modeling of Concrete Performance.* New York: E & FN Spon

Maekawa, K., Ishida, T. and Kishi, T. (2003). "Multi-scale modeling of concrete performance- integrated materials and structural mechanics", *Journal of Advanced Concrete Technology (JCI)*, 1(2): 91–126

Nakarai, K., Ishida, T. and Maekawa, K. (2006). "Modeling of calcium leaching from cement hydrates coupled with micro-pore formation", Journal of Advanced Concrete Technology, 4(3):395–407

Ouchi, M. (2007). "Cement production in Asian Countries", Journal of Civil Engineering, Japan Society of Civil Engineers, 92(5):58–59

Torrent, R. (1992). "A two-chamber vacuum cell for measuring the coefficient of permeability of air of the concrete cover on site", *Materials and Structures* 25(150): 358–365

7. Anti-seismic Design, Diagnostics and Reinforcement for Concrete Structures

Koichi Maekawa

7.1 Introduction

In the interest of having performance-based design offer more transparency to clients and taxpayers, performance assessment methods occupy a central position from the viewpoint of structural mechanics and engineering. This rational way of assuring the overall quality of infrastructures may create cost-beneficial design and construction that exactly satisfy several require-ments assigned to engineers. Life-cycle-based diagnostics of structures is required explicitly by societies and organizations. Furthermore, there is emerging a greater need to verify the remaining functionality of damaged facilities in order to extend their service life. To meet these challenges, it is desirable to formulatean explicit system for the prediction and simulation of structural life serviceability and safety under specified loads such as earthquake and ambient conditions.

In this chapter, the author proposes an integrated platform of solid mechanics and hydrothermal dynamics of materials and structures with multi-scales of referential control volume on which each physicochemical factor is applied. In more detail, a constitutive model is discussed with regard to cracking in reinforced concrete (RC) elements, and a mathematical overlay of hydrothermal state variables is presented for multiscale and multichemical mechanical coupling with soil foundation. Recent application of the multi-scale approach to practical problems is introduced for urban regeneration, and the direction of future development is discussed as an integrated knowledge-base of structural concrete and soil foundation.

Y. Fujino, T. Noguchi (eds.) *Stock Management for Sustainable Urban Regeneration*,
© 2009 to the complete printed work by Springer, except as noted. Individual authors
or their assignees retain rights to their respective contributions; reproduced by permission.

7.2 Multi-Directional Crack Mechanics

A scheme of RC modeling used for an integrated platform of both safety and life-cycle assessment is simply illustrated in Figure 7-1 (Maekawa et al. (1999, 2003)); Nakarai et al. (2005a, b)). Multidirectional cracking and its interaction are taken into account by the active crack approach (Maekawa et al. (2003)) on the smeared compression stress field (Collins and Vecchio (1982)). All microscopic physical states (cracking, yielding, crack shear slip, remaining stiffness of fractured materials) are included in the constitutive modeling. The stress-carrying mechanisms are composed of compression/tension parallel and normal to cracking and shear transfer. By the active crack method (Maekawa et al. (2003)), the primary cracking of governing nonlinearity of structural concrete is identified if some cracks intersect non-orthogonally. Here, path-dependent parameters are renewed only along the active crack in each load step of time.

For seismic analyses in time domain, the plastic localization of reinforcement is of importance for rationally simulating largely deformed elements. The spatial averaging of local stress and strain along reinforcement is applied for structural analysis with finite elements, as shown in Figure 7-2. Since the local yield occurs at the crack location and the rest of the domain generally remains elastic, the averaged stress strain relation of deformed reinforcing bars differs from that of a single bare bar. The following hardening of the element is much associated with extension of plastic zones, and the averaged hardening stiffness is computed by considering the reinforcement ratio, tensile strength of concrete and properties of reinforcing bars (Maekawa et al. (2003)). When the load reversal is produced in a single direction, near orthogonal two-way cracking is experienced. Here, the crack-to-crack mutual interaction is not so great as to consider the shear transfer of each intersecting cracks. Then, the smeared crack methods that assume coaxiality of stress and strain fields (rotating crack) may function successfully for structural analysis, and the model of shear transfer does not play a central role in mechanics.

However, the multidirectional and nonproportional loadings may create three- and four-directional cracking that intersects each other in finite element domain. When thermal and drying expansion and shrinkage would be coupled with seismic loads, principal stress directions rotate considerably. This situation tends to create multidirectionally intersecting cracking with strong interaction. Figure 7-3 shows an example of experimental verification with three- and four-directional cracking in two-way reinforced RC panels under combined in-plane shear and normal stresses. The in-plane stresses were actively controlled by the internal hydraulic-pressure torsion moment produced by a couple of

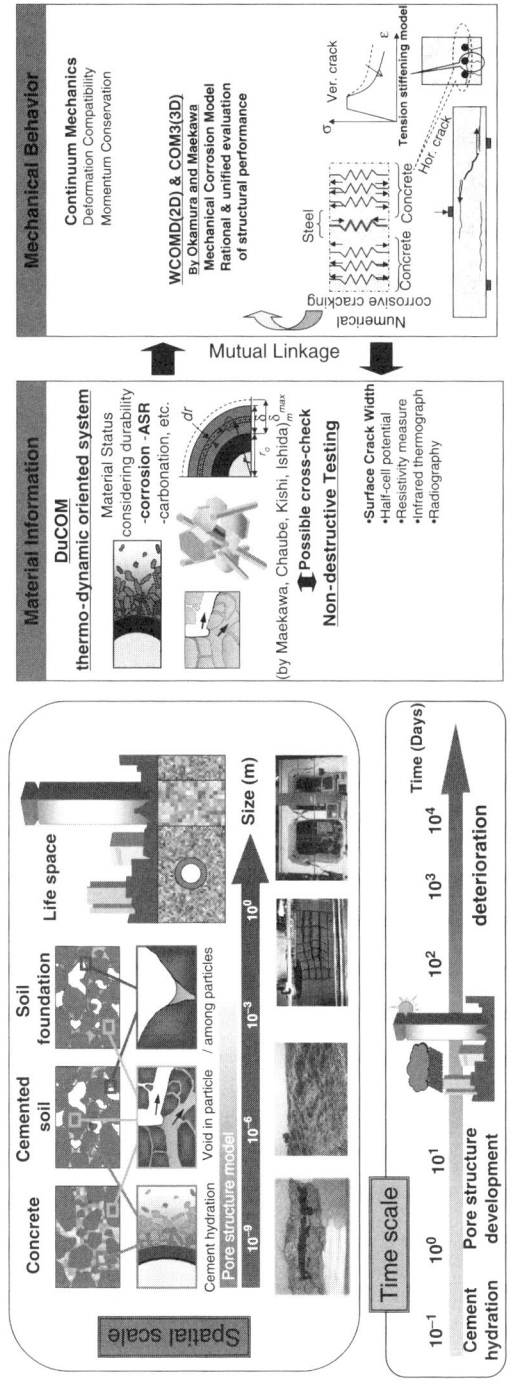

Fig. 7-1. Coupling of thermo-hydro dynamics and damage mechanics for life cycle assessment of structures with soil foundation interaction

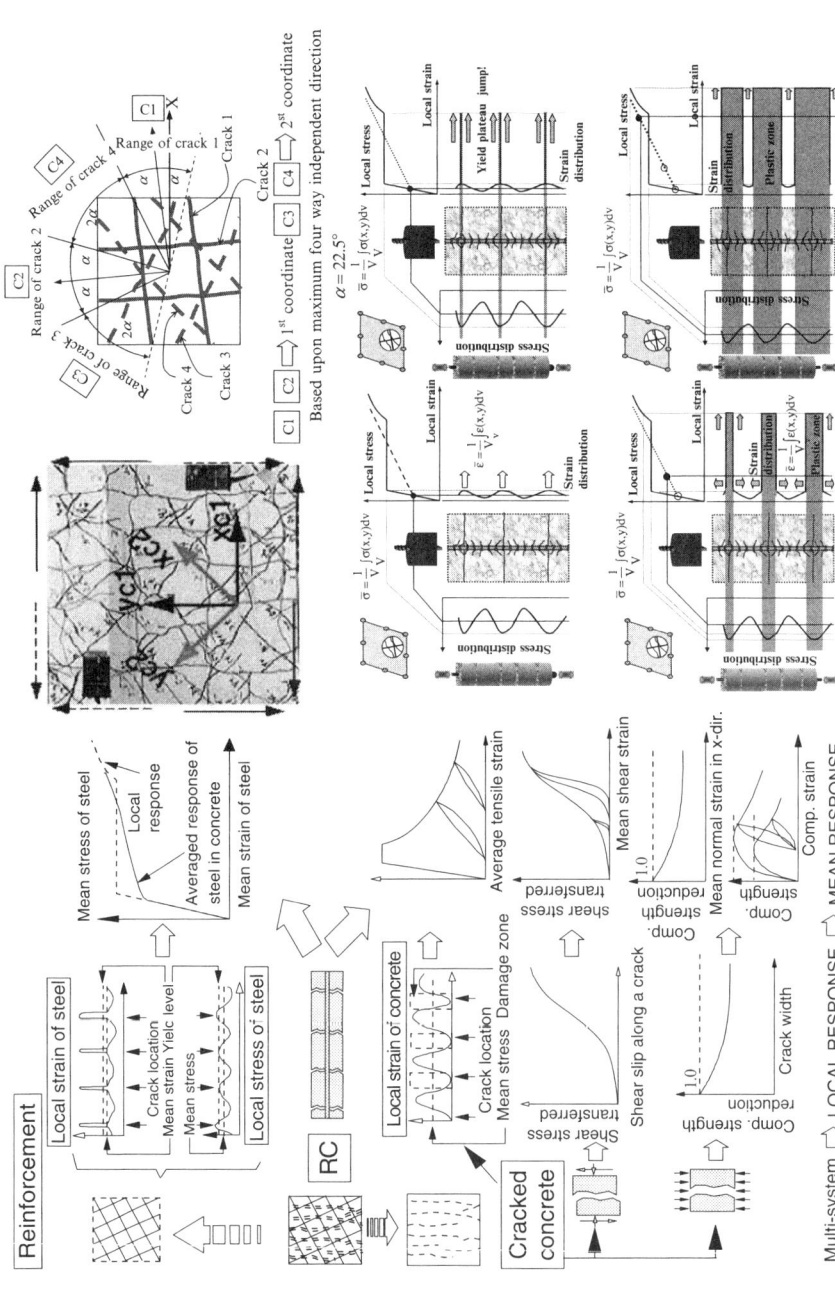

Fig. 7-2. Formulation of in-plane constitutive model with multidirectional cracking

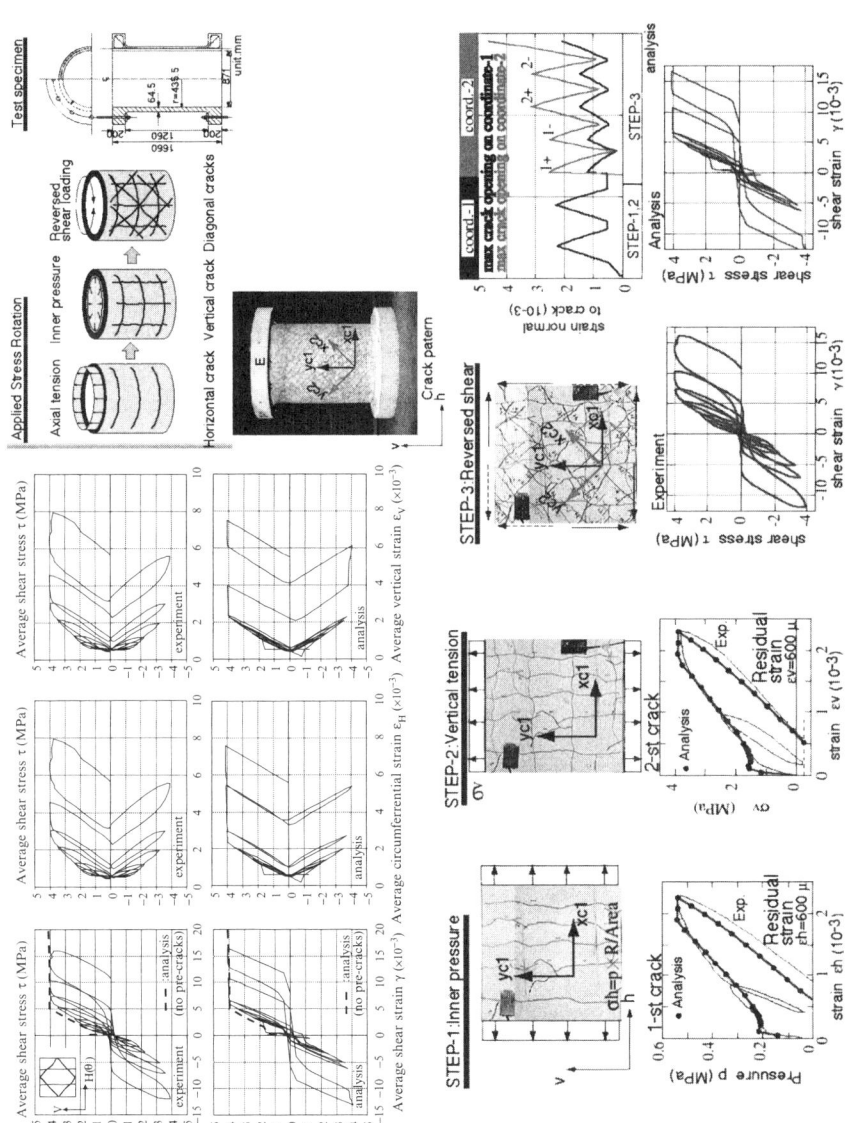

Fig. 7-3. Experimental and analytical behavior of specimens A-2 (four-way cracks)

jacks and axial compression. The nonorthogonal crack intersection frequently takes place when the principal direction of applied loads varies and/or external forces and ambient actions are coupled together.

Constitutive models have to be verified over the larger scales of structural concrete such as beams and shells, because stress states and loading paths cannot be fully reproduced only by experiments at the specimen level under rather uniform fields of stresses and strains. Shear wall experiments have been used for verification of in-plane RC modeling under monotonic as well as cyclic loads (Maekawa et al. (2003)) of small numbers of repetitions. It is recognized that in-plane RC models are well applied under both static forces and dynamic excitation. The verification has been extended to 3D shells, beam-column lineal members and closed frames (tunnel model) surrounded by soil foundation. The inelastic creep and fatigue in compression, tension and shear transfer along crack planes have been included in the computational platform (El-Kashif and Maekawa (2004)).

Figure 7-4 shows the experimental verification by using the semi-real scale mockup for tunnels and ducts. Here, careful focus was placed on the shear capacity and ductility. Specimen A, which corresponds to the seismic design code before 1980, exhibits the clear diagonal shear cracking, and it can be seen that the overall shear ductility is much degraded in both experiment and analysis. Here, the load-carrying mechanism against the dead weight of overlay soil foundation is damaged. But, Specimen B, which is well reinforced against shear failure by stirrups, shows higher ductility. Plastic hinges are produced at each corner of vertical members, but diagonal shear cracking is successfully avoided. A computational approach may fairly simulate the structural behaviors. In these analyses, the multidirectional cracking model has been examined as well by reproducing alternate change of principal stress directions in 3D spaces. For practical use of the computational platform, the multiplastic potential model for a soil skeleton is included with the underground pore water to take the effects of liquefaction into account.

The behavioral simulation of structural concrete was also examined three-dimensionally by using RC shells subjected to multidirectional loading as an analogy of real earthquake ground motions (See Figure 7-5). Three- and four-way directional cracking is introduced to box-shell components with strong mutual interaction. In order to check the versatility, a cylindrical shell was also tested for checking the ductility of underground LNG storage tanks. After these experimental verifications, global safety was quantitatively specified for the computer-based simulation in the scheme of performance-based design.

Figure 7-6 shows the soil container that includes the RC box culverts and that was set on the shaking table. The soil pressure applied on the walls of

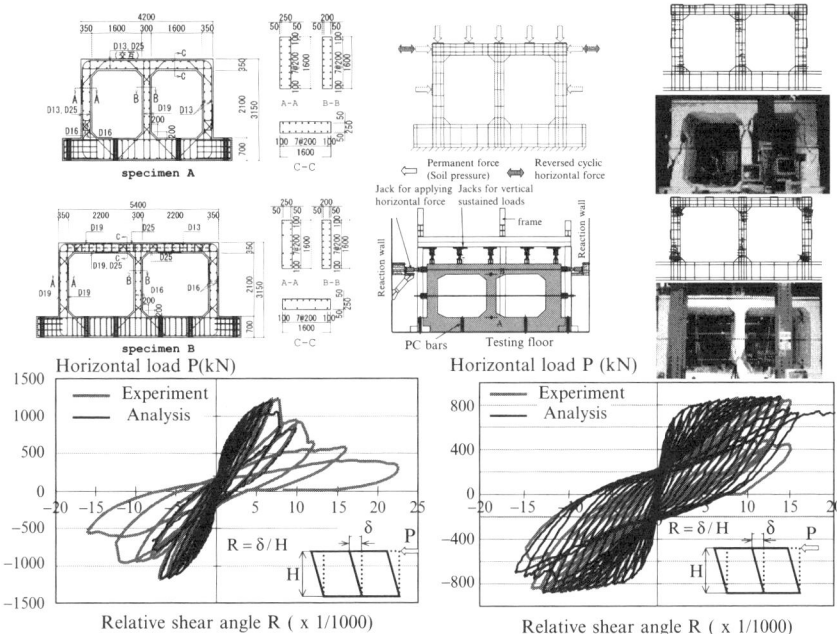

Fig. 7-4. RC underground box-vents subjected to static loads (Source: Soraoka et al. (2001))

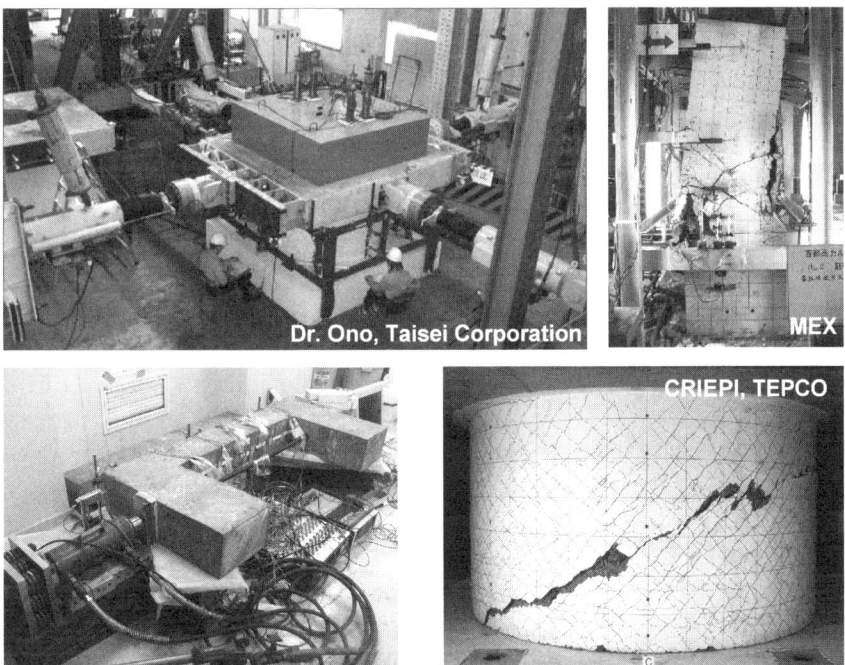

Fig. 7-5. RC shells subjected to multidirectional static loads

the embedded RC ducts and the average shear deformation of the structure was carefully compared with the analytical results. One RC duct is directly fixed on the bottom of the soil-structure system to reproduce the real boundary condition of nuclear power plants constructed above the rock foundation. Another RC is floating in the soft foundation. The computed and experimentally obtained structural deformation can be compared with each other, with the reasonable consistency as seen in Figure 7-6. As of right now, the nonlinear finite element analysis has been authorized as a tool for examining the seismic safety performance in the scheme of designing LNG storage tanks and RC aqueducts for nuclear power plant facilities for practice in Japan.

7.3 Hydrothermal Physicochemical Modeling

Stated variables of hydrothermal dynamics are further required for life-cycle assessment, especially for durability assessment related to material properties. Volumetric change caused by temperature and long-term moisture equilibrium in micropores are associated with cracking and corresponding serviceability, and corrosion of reinforcement has much to do with migration of chemicals through micropores. Thus, the coupled system as shown in Fig. 7-7 was proposed (Maekawa et al. (1999, 2003)) to simulate the entire thermomechanical state of constituent material and structures. For computing the hydrothermal equilibrium, the multi-scale analysis platform *DuCOM* (Maekawa et al. (1999, 2003)) was used. Micropore geometry and spaces are idealized by statically formulated pore distribution, and internal moisture balance is simultaneously solved with mass conservation requirement. The moisture migration and diffusivity are computed based on the micropore size distribution and the space of the condensed water channel.

Chloride-ion migration and other chemical reactions such as carbonation and calcium leaching are overlaid on this system (Maekawa et al. (2003)); Nakarai et al. (2005a)). The conductivity and diffusion characteristics for mass transport are calculated based upon computationally formed micropore structure. The computation of multiphysicochemical events is carried out by means of sequential processing with a closed-loop predictor-corrector method (Fig. 7-7) (Maekawa et al. (2003)). The temperature dependent volume change is considered as an offset strain in constitutive modeling. But, concrete shrinkage associated with microclimate in CSH gel and capillary pores is directly linked with the macroscopic constitutive model (see Section 7.2) with regard to micropore pressure and disjoining pressure originating from Van del Waals and Coulomb forces.

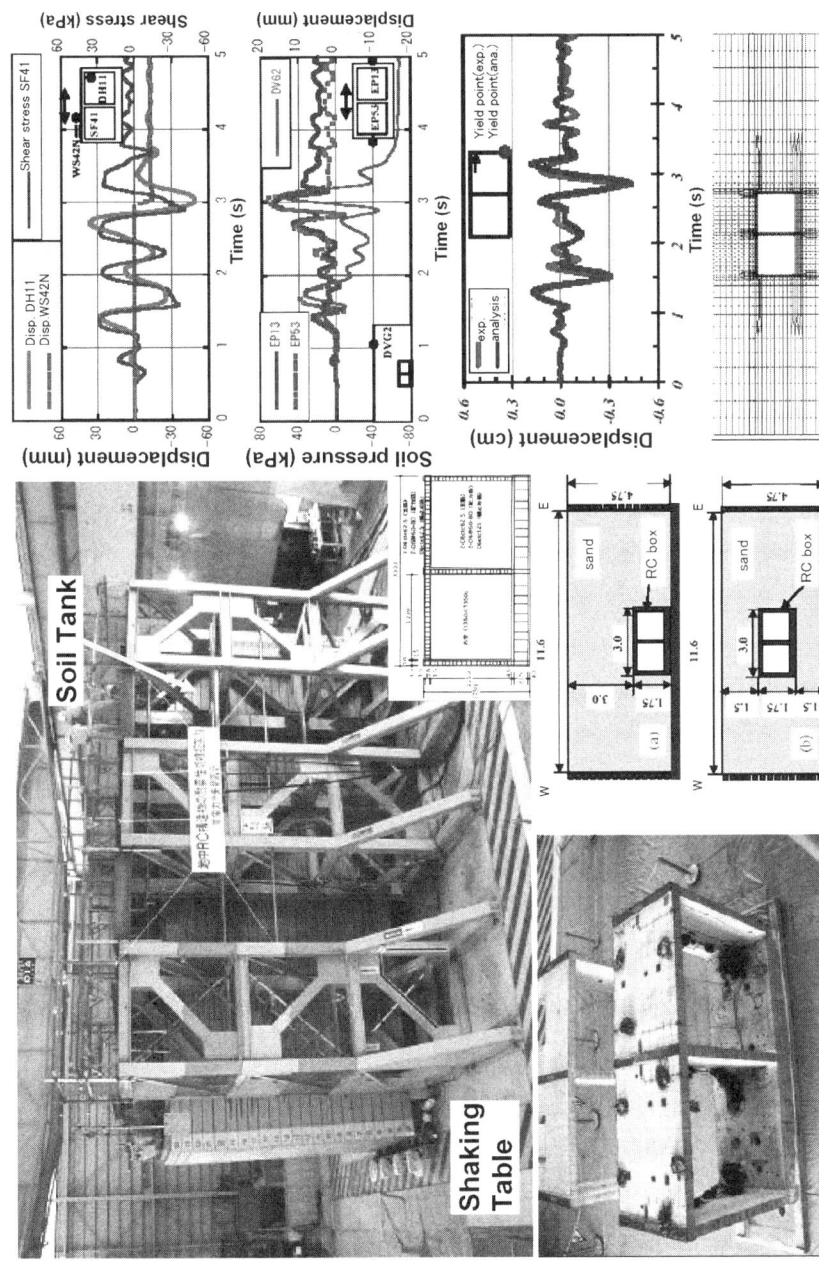

Fig. 7-6. RC underground box-vents subjected to dynamic shear (Source: JSCE (1999))

Microcorrosion rate is also computed by simulating migration of O_2-CO_2 gas and chloride ions (Maekawa et al. (2003)), and the effect of corrosion is integrated in the structural analysis (Toongoenthong and Maekawa (2005)). These thermodynamic state variables are incorporated into the constitutive modeling before cracking. In this computation, the thermodynamic equilibrium requirements are simultaneously solved, such as multi-ion balance, proton electro-balance, adsorption-desorption isotherm (see Fig. 7-7). Thus, we have approximately 230 simultaneous equations to be solved numerically for chemo-physical and mechanical behaviors of different scales.

7.4 Coupling of Damage Mechanics with Physicochemical

Cracking is also influential in mass transport of gases and dissolved ions. These cracks through which ion substances can easily migrate are mutually linked with hydrothermal dynamic analysis by the hierarchy type of multiscale modeling as shown in Fig. 7-7. This simulation can be mainly used for life-cycle assessment of structural concrete and examination of remaining functions of existing infrastructures. Cracking of concrete causes accelerated diffusion of chloride. It may allow deeper penetration of chloride and other substances. In the analysis, diffusivity of substances is regarded as a variable in terms of computed averaged strain of concrete finite elements.

Corroded steel produces volumetric expansion and results in internal self-equilibrated stress, which may lead to additional cracking around reinforcing bars. Figure 7-8 illustrates the way to amalgamate the damage mechanics and volume expansion of generated corrosion gels. The effect of corrosion gel product formation is considered in the constitutive modeling of reinforcement in the transverse direction. The noncorroded core steel and the corroded clusters with different mechanical properties are treated as a fictitious aging material of varying volumetric stiffness and expansion according to the magnitude of corrosion. This growing steel is embedded in each finite element, similar to a smeared crack approach.

If the corrosion is concentrated around the anchorage zone of main reinforcement, its structural capacity gets reduced with a different crack propagation pattern than those of sound ones (Toongoenthong and Maekawa (2005a, b)) (see Figure 7-9). The diagonal crack that reaches the bending compression zone is initiated by the corrosion crack tip created along the longitudinal main reinforcement. Finally, the diagonal crack is driven to the beam

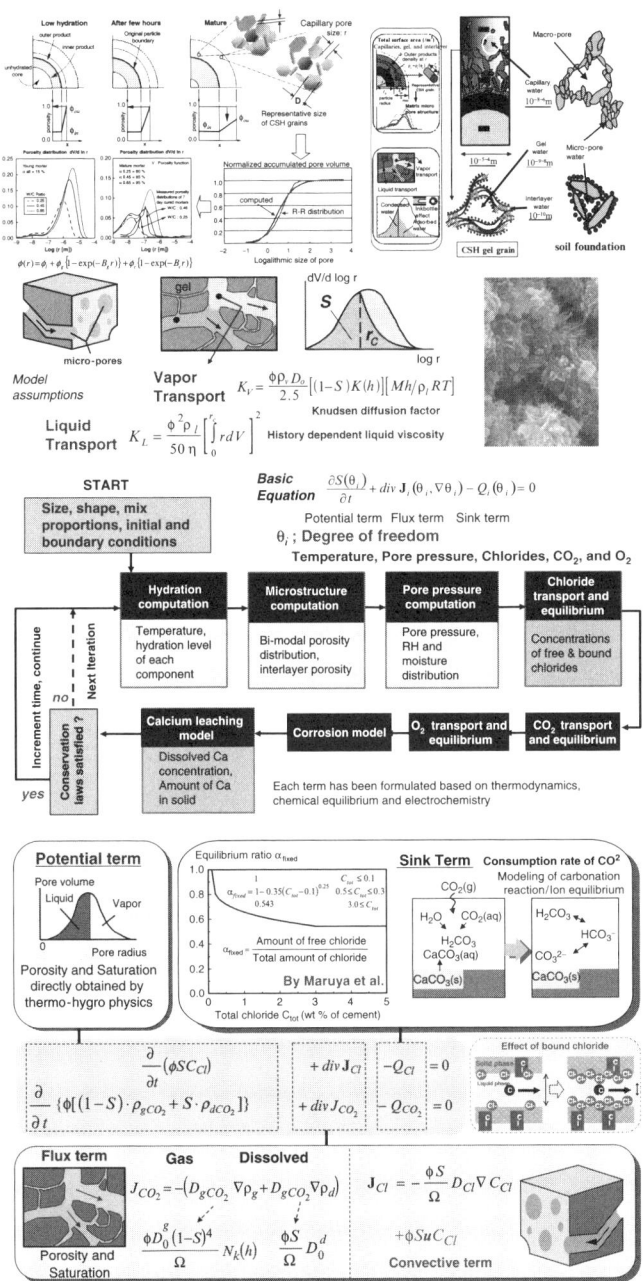

Fig. 7-7. (Top) Statistical expression of CSH microporosity and the moisture equilibrium under pore moisture potential. (middle) Flowchart of solving multiphysicochemical events. (bottom) Mircomodeling of CSH gel and capillary pores and multiphysicochemics

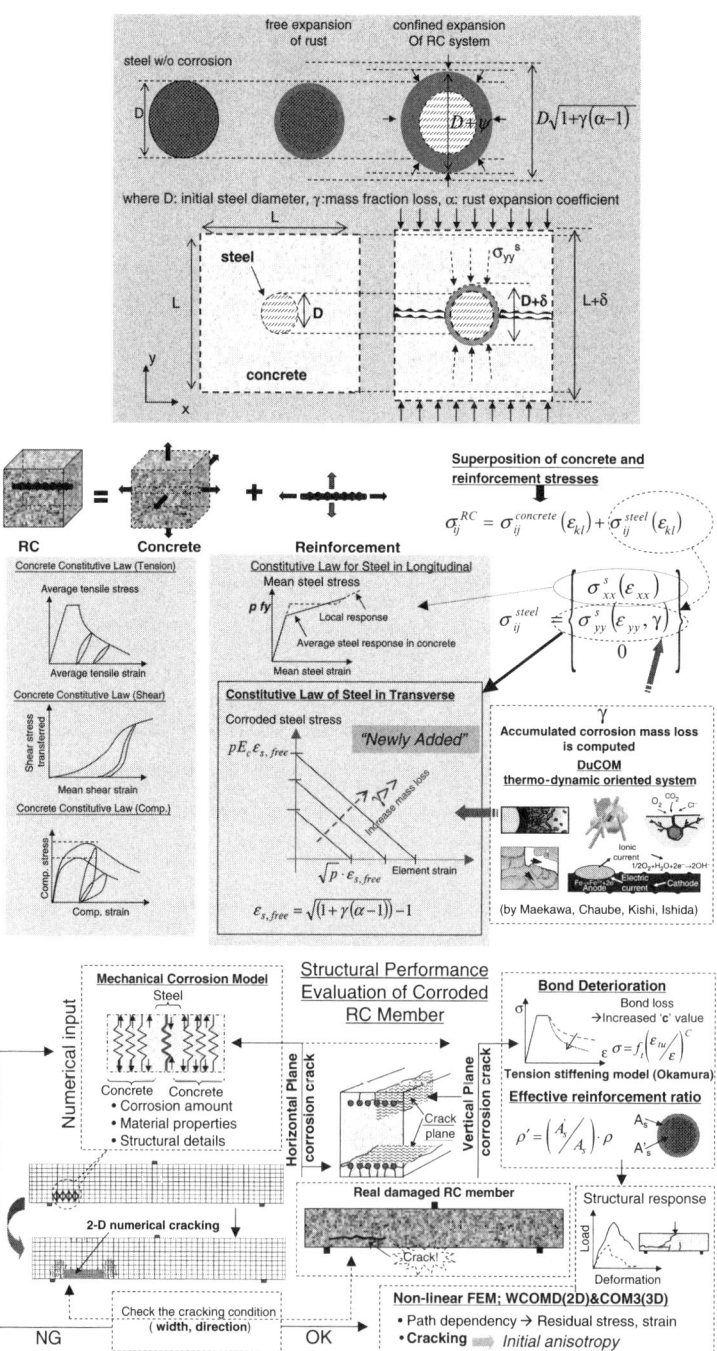

Fig. 7-8. Simulation of corrosion of r/f bars and structural performance assessment

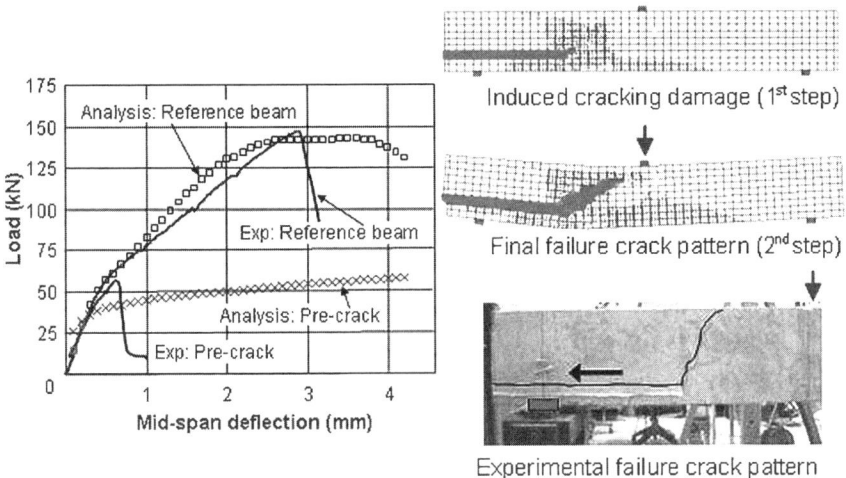

Fig. 7-9. Analysis result of beam with inherent crack-like defect till the anchorage zone

support. Apparently, the localized corrosion is seen to deteriorate the anchorage performance of longitudinal reinforcement. The acceleration test of corrosion of steel in a RC beam by galvanostatic charge also substantiated this simulation result.

When corrosion cracking develops over the beam, shear safety performance differs from the nondamaged reference case (Satoh et al. (2003)). Figure 7-10 shows load-displacement relations for a RC nondamaged reference and corroded specimen, which was submerged into a sodium pond for accelerated corrosion. The result here produced was uniformly distributed corrosion along the whole longitudinal steel (2.1% as the mass loss).

Main reinforcing bars were bent up 90 degree inside the anchorage zone. Thus, comparatively satisfactory anchorage capacity is expected. In this case, the stiffness of the beam is much reduced but the capacity is a bit increased. The macroscopic bond loss in the shear span leads to retarded propagation of diagonal shear cracking and may elevate the shear capacity. Computation can capture this property.

Figure 7-11 shows the corrosion crack propagation in experiment and simulation. The corrosive mass loss can be computed by *DuCOM* under the constant chloride concentration on the surface (see Section 7.3). The

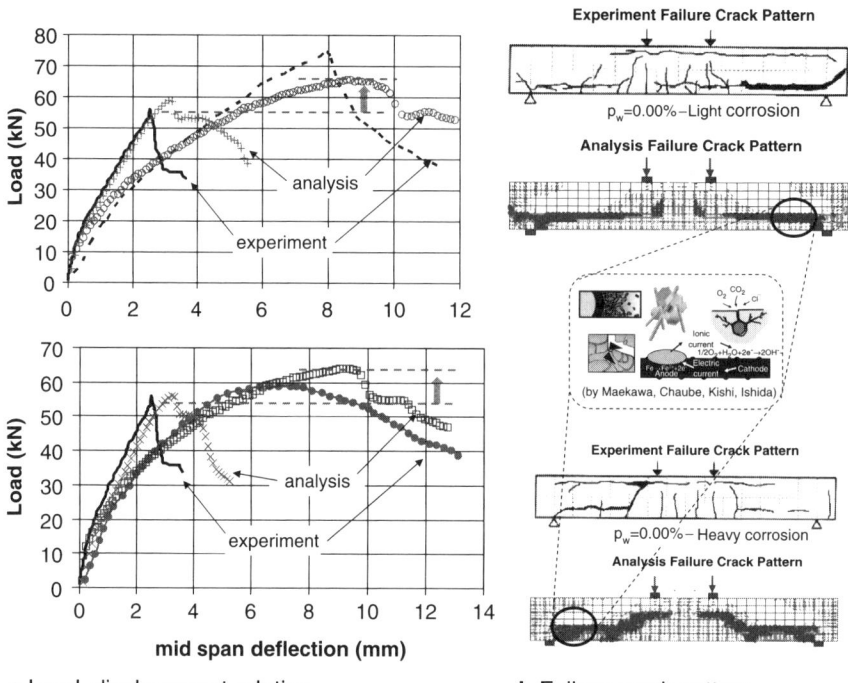

a Load–displacement relation   **b** Failure crack pattern

Fig. 7-10. Simulated shear capacity and cracking of corrosion beam

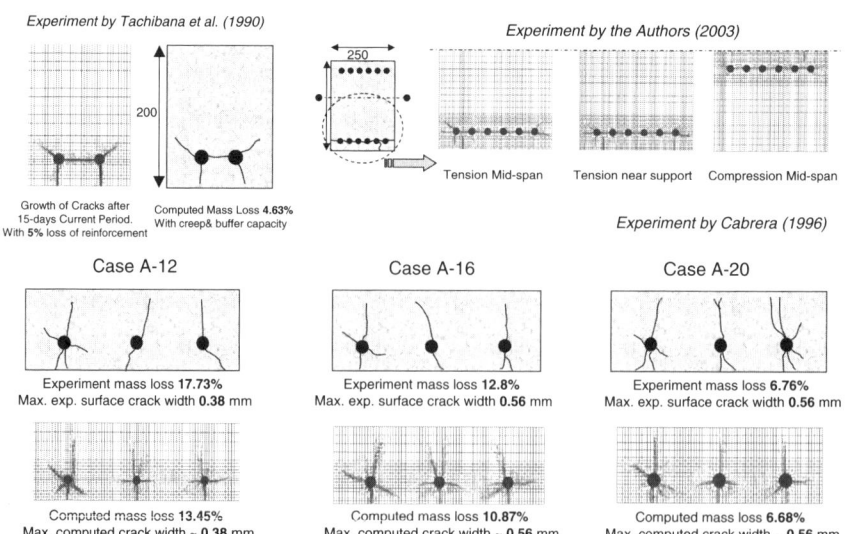

Fig. 7-11. Simulated shear capacity and cracking of corrosion beam (Source: Toongoenthong and Maekawa (2005))

corrosion gel product is assumed to have been created around the mother steel bars, and the transverse stress normal to the reinforcement axis is computed as shown in Figure 7-8. The crack patterns of the same surface crack width and the corresponding corrosion mass loss are compared with each other. The crack orientation and ligaments are simulated fairly. In these analyses, the injection of corrosion products via colloidal gels into crack gaps was taken into account. Otherwise, the rational simulation of corrosion cracking and the life of concrete cover may not be achieved.

Figure 7-12 shows the fatigue simulation of RC beams under high-cycle shear. When a precorrosion crack develops around the anchorage zone of the web reinforced beam, the degradation of member stiffness is much reduced and large deflection is produced even though the static capacity is almost the same as that of the nondamaged one. The average strain of web reinforcement for the initially damaged beam is computed to evolve several times greater than in the case of a sound structure. In this case, a clear mode of collapse in shear cannot be seen, but the large slip along the main reinforcement proceeds at the anchorage because of confinement by the web steel. When the fatigue life is tentatively defined in terms of the localized shear slip along the main reinforcement, the apparent S-N diagram shifts from the original curve as shown in Figure 7-12.

To the contrary, the precorroded cracking inside the shear span results in much longer fatigue life to non-web-reinforced members due to the loss of macroscopic bond. The corroded cracks along the main reinforcement result in an arch action against shear formed between the supports and the loading point. As far as the member stiffness is concerned, the effect of precracking is similar to the case of beams with web reinforcement. It can be understood that the corroded damage changes the load-carrying mechanism in regard to the bond between web concrete and the layer of main reinforcement. For this situation, the best way of repair is only to prevent the further evolution of steel corrosion. The recovery of macroscopic bond is not the first priority. This fact has also been experimentally verified through the use of actually damaged RC slabs for railway infrastructures. In both cases, with and without pre-induced damages, the shear failure is identical because sharp diagonal crack localization appears since no steel confinement is supplied for both cases. It should be noted that the upgraded fatigue life is observed for those elements which may fail in the mode of shear. Provided that the structural concrete is well designed to have flexural mode of failure prior to shear, the effect of corrosion on the structural performance will be opposite.

Autogenous and drying shrinkage, which can be computed by solving the moisture migration under ambient conditions (Mabrouk et al. (2004)),

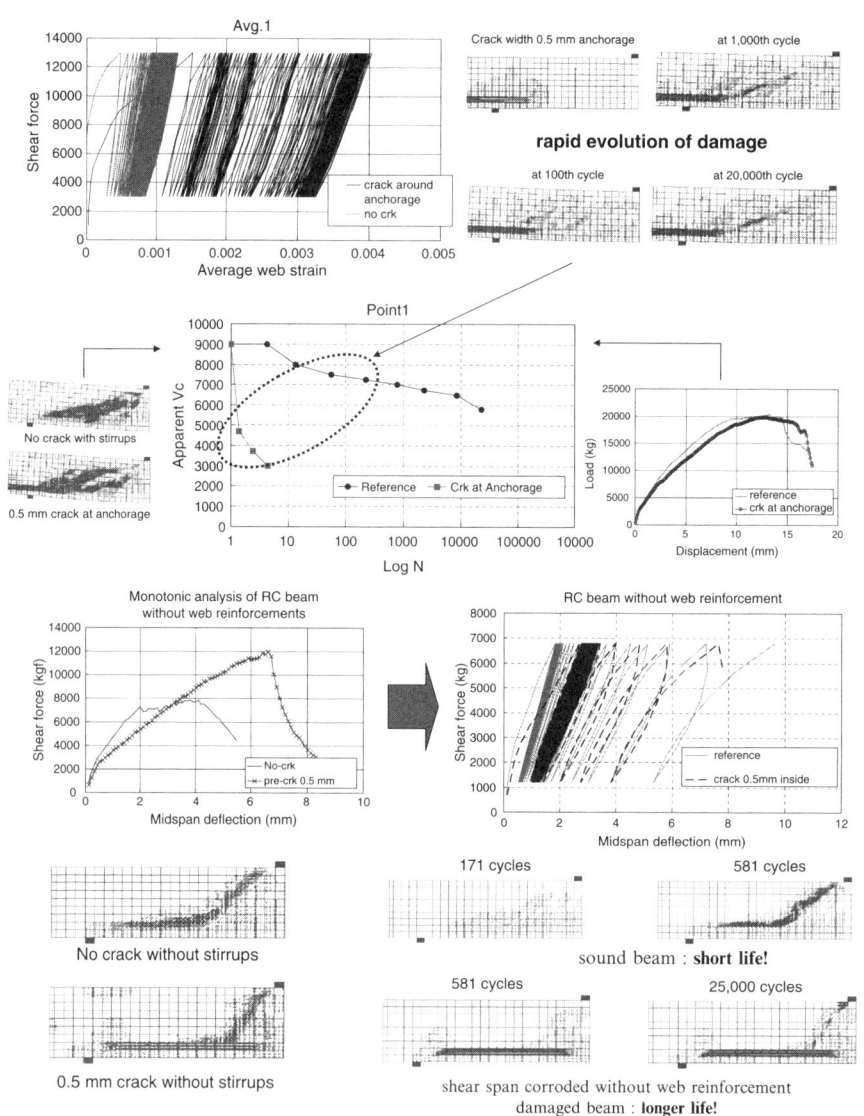

This is also to be quantitatively discussed further.

Fig. 7-12. Simulated shear capacity and cracking of corrosion beam under fatigue loads

can be directly included in the constitutive modeling of concrete in each finite element as shown in Fig. 7-13. As the moisture profile is not uniform, self-equilibrated tension is induced to concrete close to any free surfaces exposed to open air even if the member is subjected to no external forces.

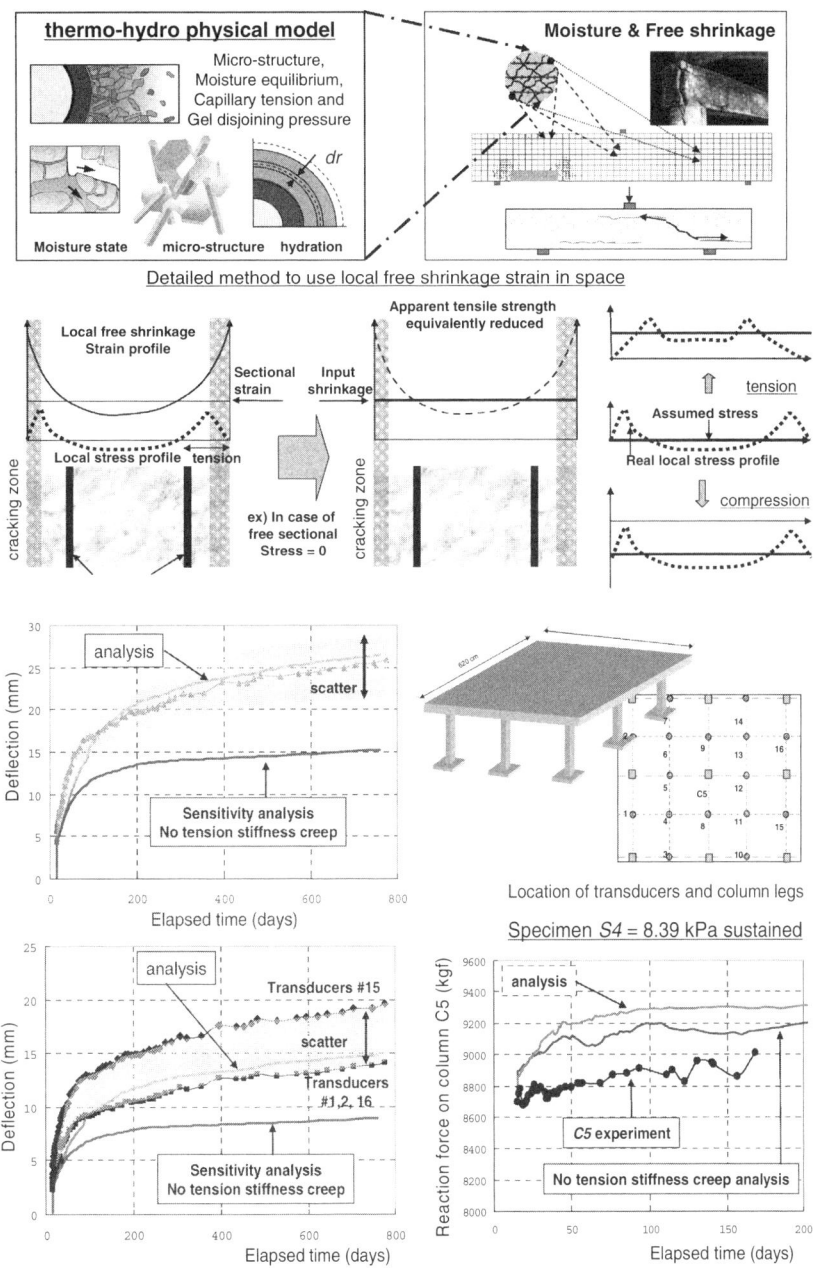

Fig. 7-13. Coupled moisture migration and time-dependent deformation of RC slabs under sustained loads (Experiment by Guo and Gilbert (2002))

The preinduced tension causes early cracking against the external sustained loads. The coupling of concrete creep in compression, shrinkage and post-cracking time-dependent tension-stiffness models yield consistent behavioral simulation with reasonable accuracy for long-term deflections of RC slabs supported by some columns as shown in Fig. 7-13. If only concrete creep in compression and tension before cracking is taken into account, the computed slab deflection does not match the reality, especially at the longer period of loading. Postcracking time-dependent modeling plays a substantial role, too.

Figure 7-14 shows the application examples for the seismic performance assessment of underground LNG storage tanks, existing RC-steel composite piers to support a long-span bridge in service, and urban in-ground and underground transportation infrastructures in the capital city of Tokyo. Multi-chemo-mechanistic nonlinear finite element analyses bring about cost benefits and behavioral simulation of structures of geometrical complexity with soil interaction. In these cases, there is no standard procedure of design to simply decide shapes of constituent structural members, dimensions, and boundary conditions. Then, the performance assessment for safety becomes crucial for proposed dimensioning and details.

The large-scale pier was found to probably fail in a sudden shear mode of failure before yield of main reinforcement. In fact, the nominal shear strength is apparently reduced according to increased sizes. This size effect was not known nor taken into account when designed. The computational approach can automatically take into account this phenomenon without special care. The shear failure of wall members of underground ducts and tunnels places many demands on engineers, because just the interior plane is accessible for strengthening. Then, the computational performance assessment is expected to possibly create some acceptable solutions. Otherwise, a huge fund for repair of underground infrastructures has to be prepared.

Figure 7-15 shows a recent example of such application to the engineering assessment for a 100-year-old railway bridge in Tokyo Metropolis (Sogano et al. (2001)). Due to uneven settlement of the foundation caused by variations in underground water level and the rapid urbanization of the 1960s, some initial damage inside the assembled brick masonry was found in the form of cracking in old bricks. Some decades ago, arch ribs were strengthened by additional RC arches inside the layer. The historical process of structures was assumed and the corresponding mechanistic actions were re-produced in the computational environment.

Afterwards, seismic ground motion was computationally applied to the numerically aged structural concrete, and the computed response was used for safety and serviceability assessment in practice. The seismic

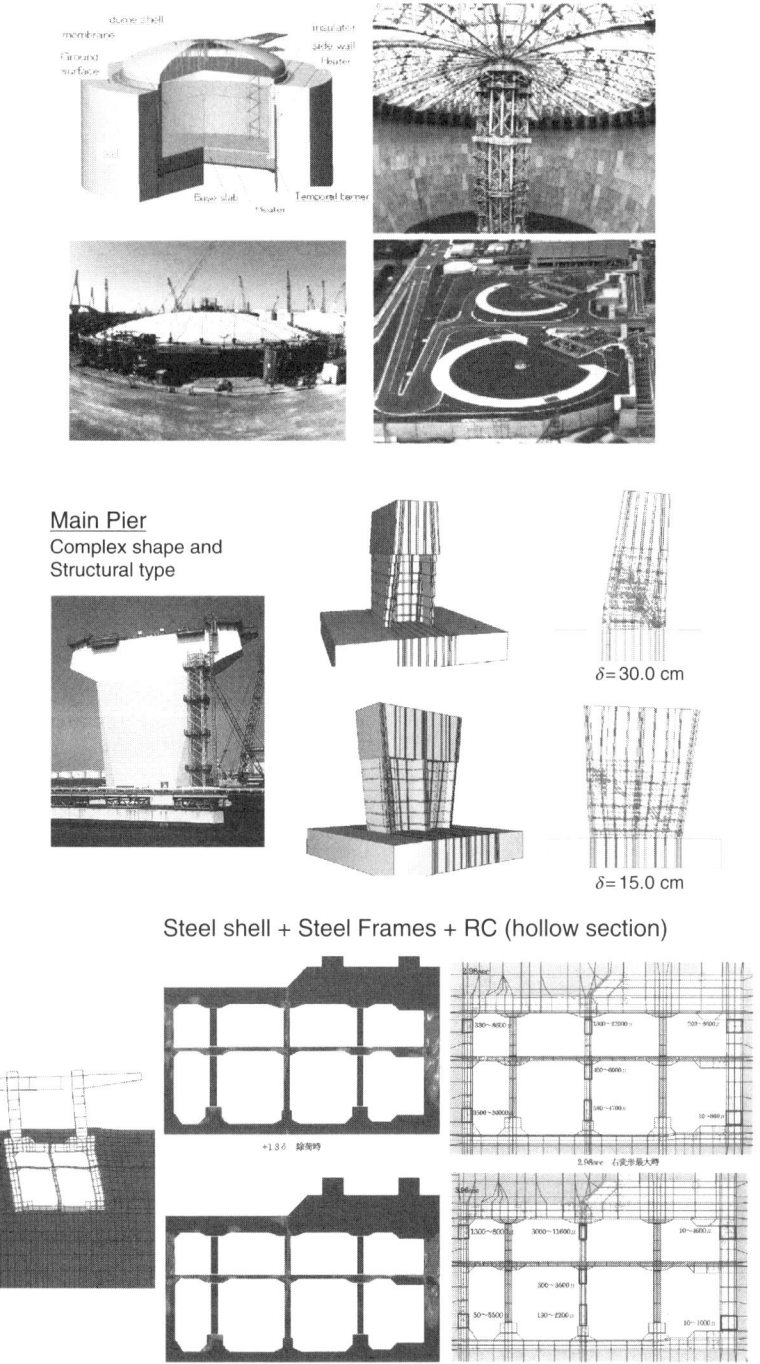

Fig. 7-14. Safety assessment of transportation infrastructures against earthquake

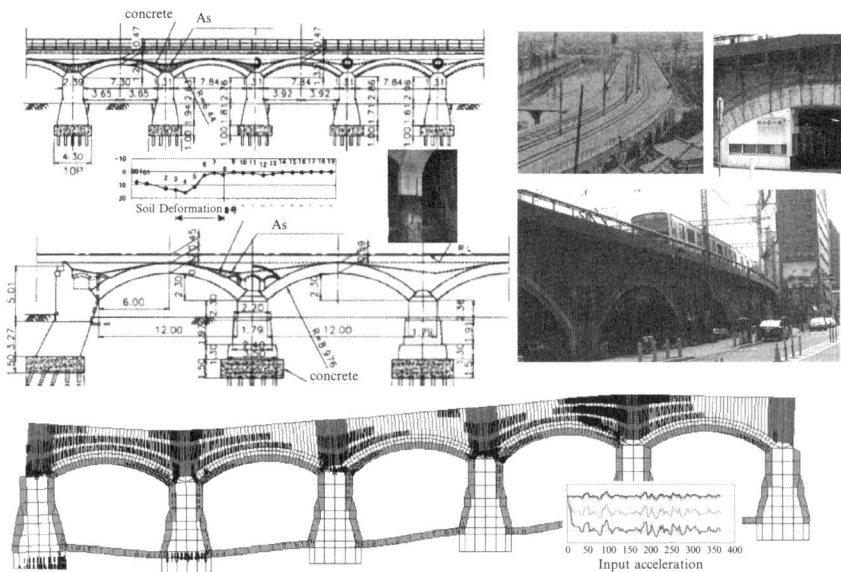

Fig. 7-15. Safety assessment of 100 year-old railway bridges (Source: Sugano et al. (2001))

performance was numerically investigated and the sustainable life of the structure with light retrofit was assessed. Long-term durability assessment and associated damage have much to do with seismic engineering issues. Now, the holistic approach is indispensable for maintenance of urban infrastructures.

Figure 7-16 shows the analysis for remaining structural safety performance of ASR-damaged RC bridge piers at a site in the west of Japan. As many of the reinforcing bars are surprisingly ruptured at the inside corners of bent portions, the anchorage performance of web reinforcement is thought to be deteriorated. It is clear that the capacity predictive formula in design codes cannot be applied here, because these design tools and knowledge have been developed based on the assumption that structural details related to hook, splices, bent radius, concrete cover, etc. have been satisfied. At the same time, the effect of dispersed cracking and self-equilibrated prestressing caused by ASR expansion has to be considered. Inescapably, the full nonlinear structural simulation is just the tool needed for calculating the remaining capacity in consideration of anchorage deterioration of stirrups and cross-sectional loss of main reinforcement.

As a matter of fact, the computed capacity of ASR-damaged members was predicted to gradually increase in accordance with the magnitude of ASR expansion. This is attributed to the prestressing effect and self-equilibrated

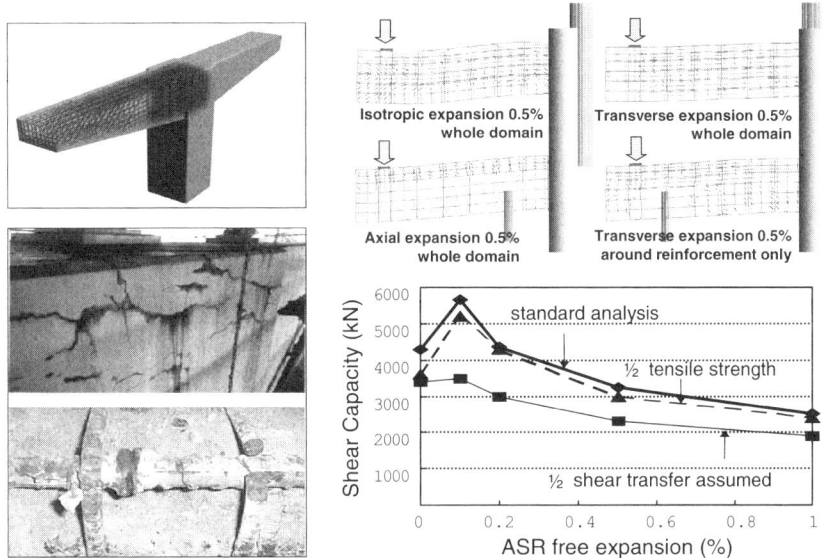

Fig. 7-16. ASR-damaged RC bridge pier and capacity simulation (Source: JSCE (2005a))

compressive axial force that elevates the shear capacity. If an excessive expansion beyond the yield of reinforcement is computationally assumed as shown in Figure 7-16, the capacity starts to decline together with the structural stiffness. Thus, the way of strengthening and/or repair must be different according to the induced level of expansion. In this case, the effect of steel rupture at the extreme ends of reinforcement is comparatively little, because the bond deterioration of reinforcement is fortunately small by volume compared to the size of the damaged structural members.

Figure 7-17 shows the PRC bridge in which plenty of shrinkage cracks were unexpectedly induced to the main viaducts due to excessive shrinkage of concrete and unnecessarily heavy reinforcement in terms of serviceability limit states. These cracks penetrate through whole sections of damaged members. The actual compliance of the real viaduct in each span was reported to be much greater than the design value. Here, both fatigue life and the substantial safety of the bridge system in terms of future earthquakes were questioned. The JSCE concrete committee (2005) investigated the detailed damage and corresponding remaining fatigue life by using the coupled chemo-mechanical simulation. For verification of the analysis method, the design live load (1,500 kN) was applied on the deck and the incremental deflection was measured as shown in Fig. 7-17. The simulation was reported to be closer to the reality of the damaged PRC bridge.

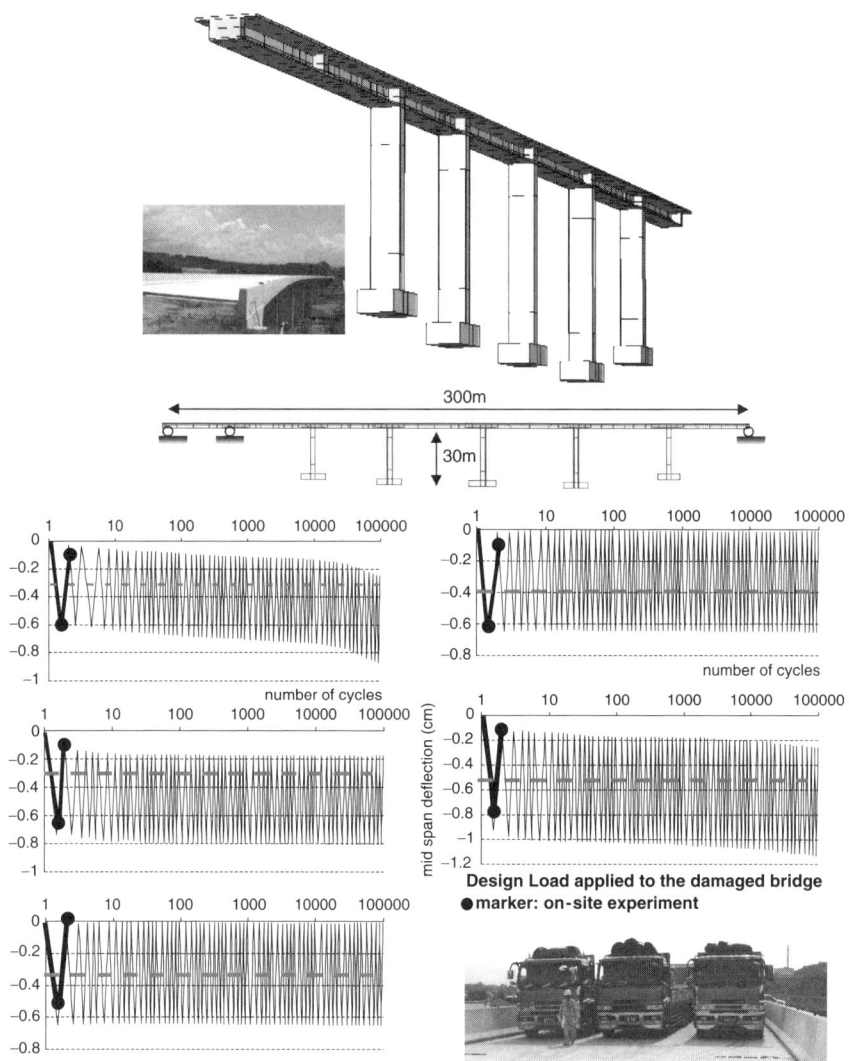

Fig. 7-17. Shrinkage cracking and RC bridge pier and fatigue simulation (JSCE2005b)

Recently, the technology for simulating nonlinear seismic responses of structures was extended to the structural concrete simulation under high-cycle fatigue loads. Two types of load, cyclic fixed point and traveling wheel-type loading, are applied to allow for a comparison of failure mode and fatigue life as shown in Figure 7-18, where the magnified displacement profile is drawn as inverted for ease of understanding. Under the fixed-point

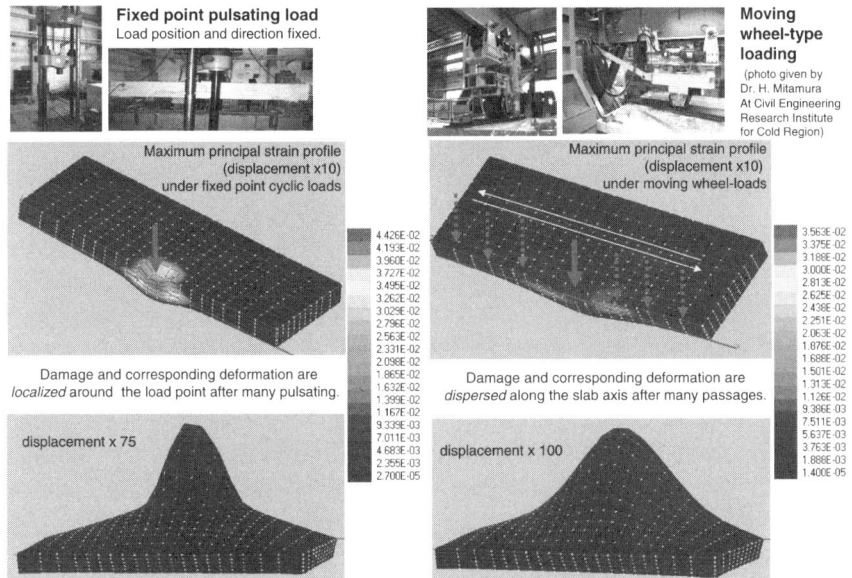

Magnified vertical displacement of the bottom face of the slab (upside down drawing)

Fig.7-18. Fatigue simulation of RC bridge deck under fixed piont pulsation and wheel-type moving load

pulsation, the deformational mode exhibits the typical punching shear failure accompanying localized conical failure planes of 3D extent.

There is a great difference in failure modes between the two types of loading. In the case of moving loads, a diagonal shear failure plane appears only in the transverse direction normal to the wheel-track, but typical shear failure modes are not in evidence in the longitudinal direction of the slab as shown in Figure 7-19. This computational simulation is very consistent with experimental observations. Figure 7-19 also show cross sections obtained after the experiment by cutting the real RC slabs.

Under moving loads, multi-directional flexural cracks occur over the whole domain of the RC slab. Diagonal shear cracks in the longitudinal direction are prevented by crack-to-crack interaction (Pimanmas and Maekawa (2001)). As a result, the load-carrying mechanism evolves from punching shear to semi in-plane load-carrying over the transverse direction. The three-dimensional direct path-integral scheme of analysis makes it possible to reproduce the mechanistic character of failure modes and deformation. Accordingly, the fatigue life is not necessarily invariant but can differ with loading and boundary conditions.

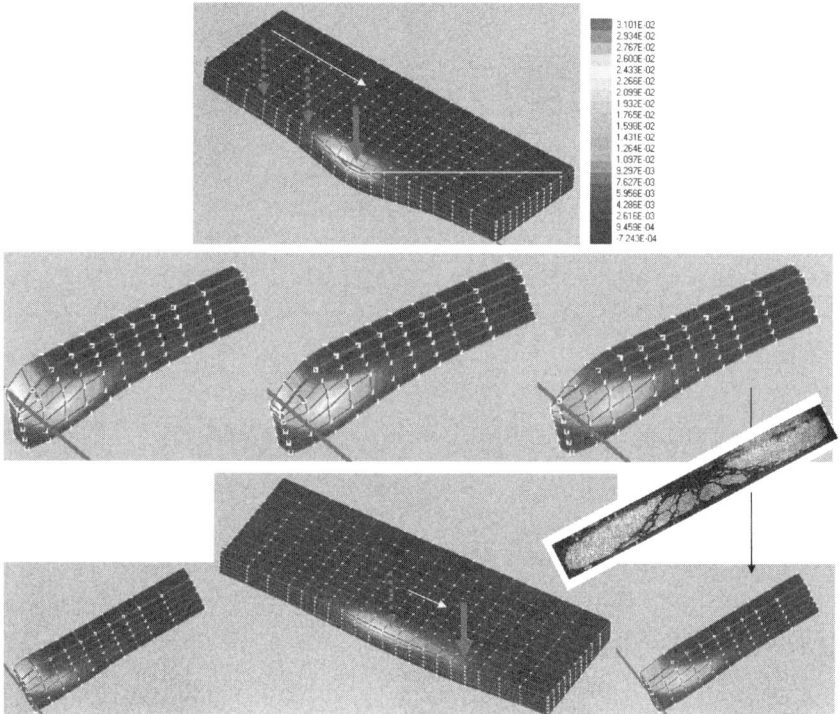

Fig. 7-19. Failure mode of damaged RC slab under wheel-type moving loads

The drastically shortened fatigue life observed under the wheel-traveling moving loading, as shown in Figure 7-20, is particularly noteworthy. Under fixed-point loading, the fatigue life diagram (S-N diagram) is quite similar to that of RC beams (Kakuta and Fujita (1982)) but with wheel-type loading the slab life-cycle is considerably degraded as known from experiments (Matsui et al. (1984, 1987)); (Perdikaris and Beim (1988)); (Perdikaris et al. (1989)). Roughly speaking, the fatigue life is reduced by $1/100$–$1/1,000$ in the computations, consistent with experimental observations. As explained by Maeda and Matsui (1984), the load-carrying mechanism may evolve as a result of crack-to-crack interaction and this results in a reduced area of diagonal shear fracture planes. This shift in the S-N diagram is well predicted by the full-three dimensional fatigue analysis, in which load cycle at failure is judged by monitoring the mode of out-of-plane deformation and deflection at the center of the slab. Fatigue shear failure is recognized as having occurred when a diagonal crack continues to expand but deformation of the surrounding block elements indicates recovery (shear localization).

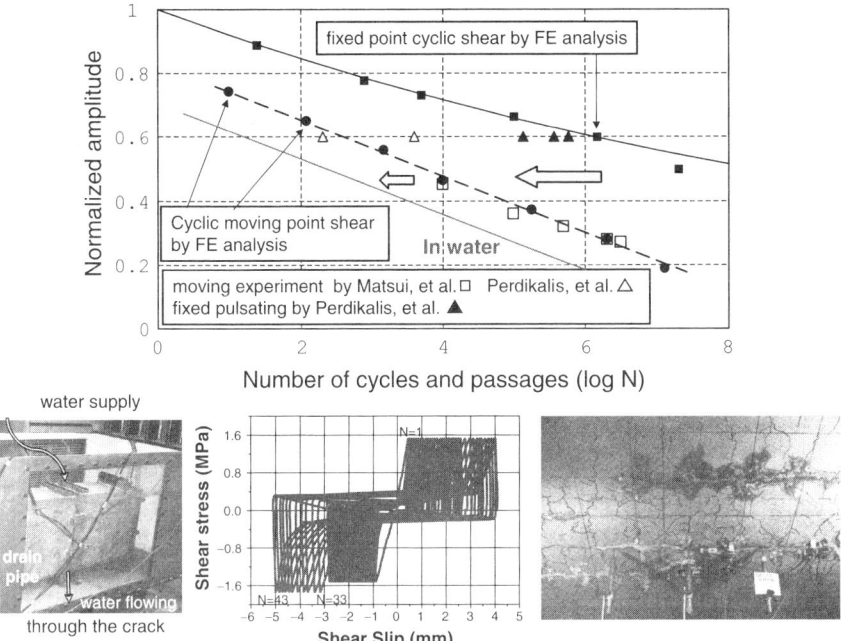

Fig. 7-20. Failure life and effect of moisture

Here, let us consider the coupling with water that exists in natural weathering conditions. The shear transfer mechanism along a crack plane is known to be seriously degraded in water. The shear transfer stiffness and capacity were computationally reduced in accordance with the experimental facts, and fatigue analysis was conducted again as shown in Figure 7-20. As a result, the fatigue life of RC slabs is further shortened, very similar to the experience observed in past decades in practice. The coupling of thermodynamic hydrophysics with structural mechanics is a crucial strategy for dealing with this urban infra-stock problem as well.

7.5 Coupling with Soil-Foundation System

The concept of mechanical nonlinear interaction of soil foundation and concrete structures was introduced in the previous section. With regard to physicochemical events, the multiscale modeling of cementitious composites can be extended to soil foundation with large-scale pores of strong connectivity as well. Figure 7-21 summarizes the micropore distribution

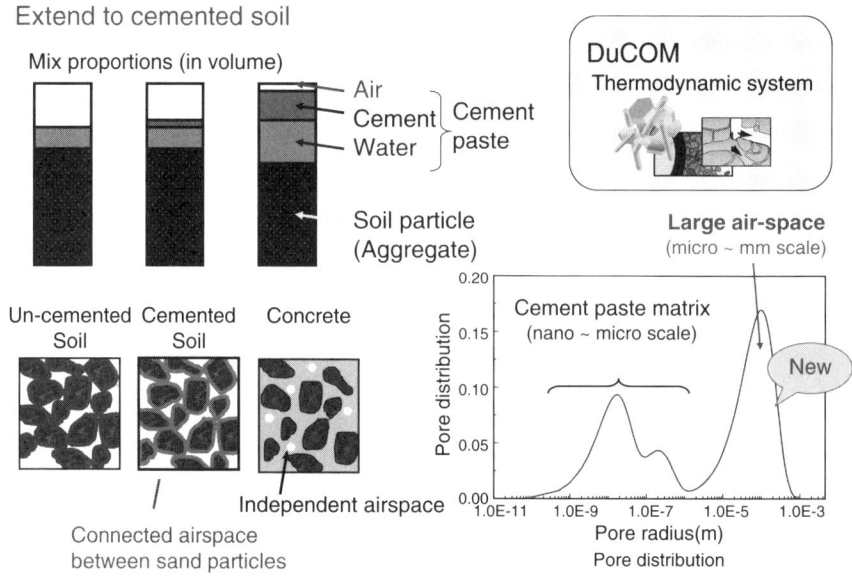

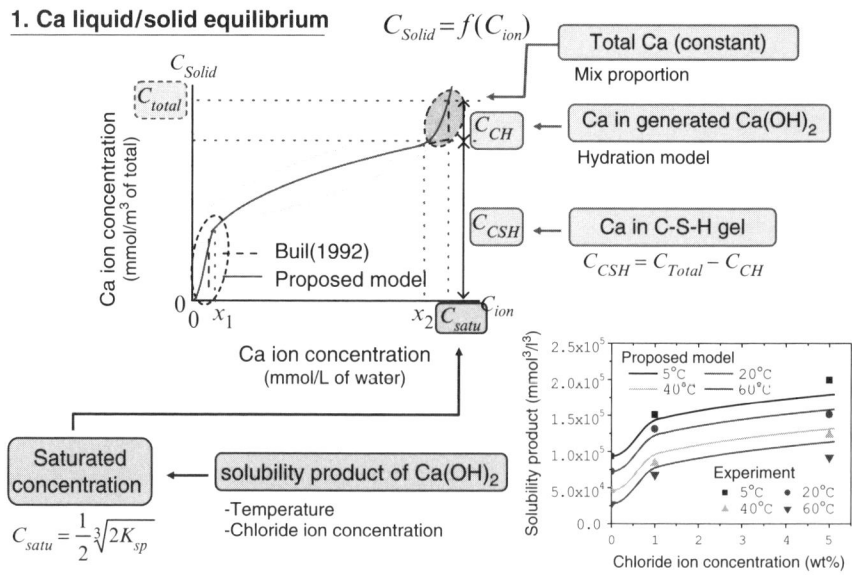

Fig. 7-21. Extension of micropore model and calcium leaching from CSH to soils

modeling. The mass transport modeling through micropores was extended to the large-scale pores in soil grains (Nakarai et al. (2005b)). The extended system is expected to be applied to so-called cemented soils. As a matter of fact, the borderline between soil and cementitious composites is disappearing from an engineering viewpoint.

Verification of the modeling was conducted in the process of experimentally obtained permeability. Leaching and mass transport of calcium ions were factored into this extended computational system for estimating extremely long-term performance of underground concrete structures like nuclear waste depositories as well as those created to deal with some environmental issues of underground water. Figure 7-21 shows an isothermal equilibrium of the total calcium concentration in all phases and the ionized one in micropore solution.

This extended simulation method can also apply to the life-cycle assessment of cementitious soil. Figure 7-22 shows analytical and experimental results of calcium leaching from the cemented sand. Leaching is associated with permeability and bulk motion of water, whose characteristic is greatly influenced by the micropore structure. This is not a given material constant but a computed value in the multi-physicochemical scheme. As the calcium exists in the form of CSH gels and $Ca(OH)_2$ in concrete, the isothermal

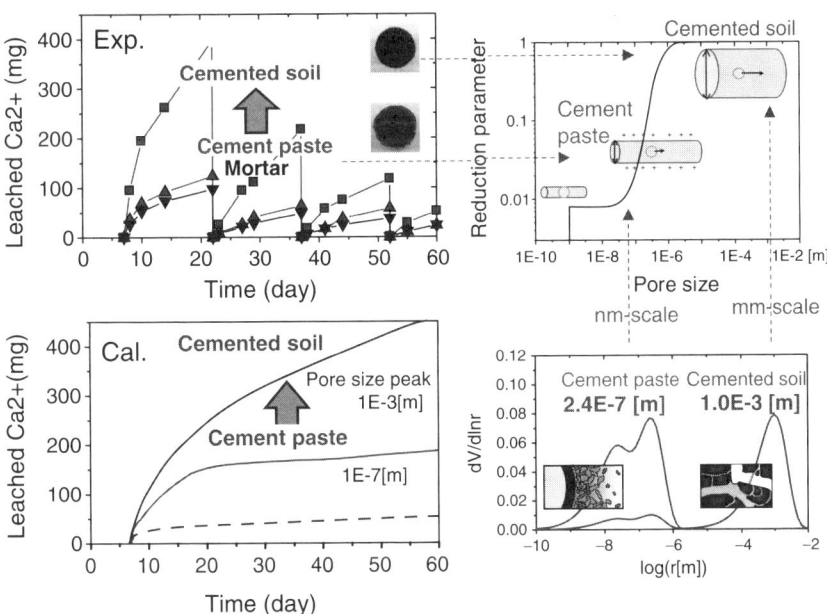

Fig. 7-22. Calcium leaching of cemented soil

curve is rather complex. In the leaching process based upon the isotherm as shown in Fig. 7-21, calcium initially dissolves from Ca(OH)₂ and subsequently leaches from CSH gels. This isothermal curve is dependent on the mix proportion of concrete and chemical characters of cementitious powers. When we use a pozzolan such as fly ash or blast furnace slag, the amount of Ca(OH)₂ is decreased and the isotherm is changed.

Figure 7-23 shows an analysis and verification of calcium leaching from underground concrete structures into the soil environment. Different boundary conditions are assumed: (1) exposed to water with constant concentration of calcium, as in the sea; and (2) soil foundation with no bulk motion of underground water. The degree of deterioration obtained from the coupled analysis with soil foundation is much lower, and as a matter of fact, closer to reality than in the analysis case, which does not include the surrounding ground. The soil foundation keeps the underground water inside its pores and the dissolved calcium ion cannot freely migrate through the micropores by diffusion. At that point, the mild profile of calcium ion concentration is realized. This is to reduce the molecular diffusion of calcium from the hardened solids. The soil ground works as a protector of concrete against leaching. The results indicate that the interaction with the soil foundation is critical for any assessment of leaching of an underground structure.

Figure 7-24 describes the outline of the modeling for calcium ion behavior in bentonite in order to investigate the influence of surrounding

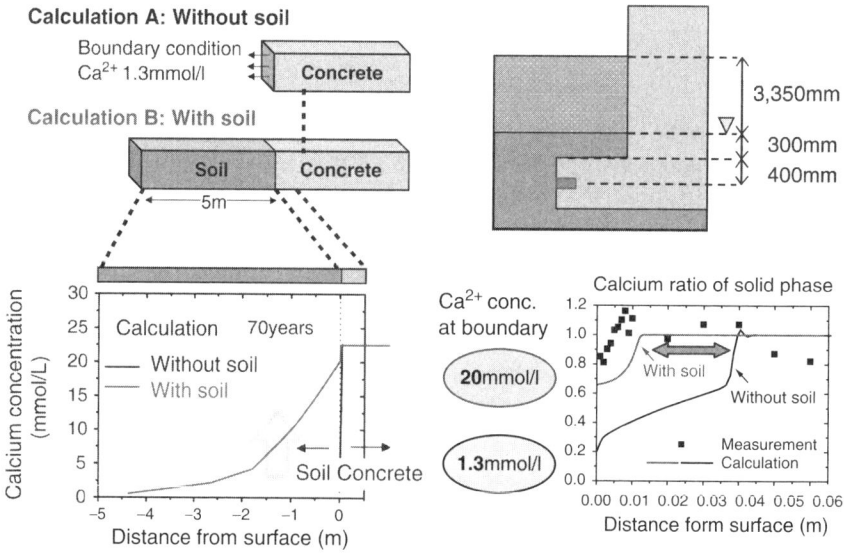

Fig. 7-23. Verification of calcium leaching from concrete

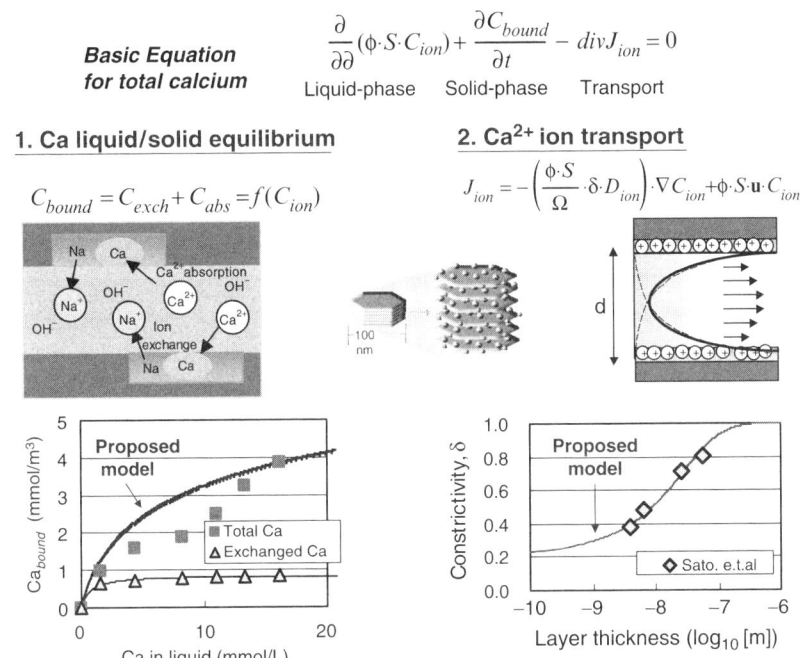

Fig. 7-24. Extended calcium ion model for bentonite

bentonite on the concrete barrier. The bentonite layer is expected to prevent moisture migration as an isolator for a nuclear waste depository. The model consists of the calcium liquid-bound equilibrium model and the calcium ion-transport model. The equilibrium model treats the binding of the calcium ion caused by the dislocation of sodium ions in montmorillonite (and the calcium ion electrically absorbed on the surface of the bentonite) as a bound calcium ion. The relation between bound and liquid calcium was determined based on the results of a simplified experiment in which bentonite particles were soaked in calcium oxide solution. In the ion-transport model, the constrictivity was taken into account as a reduction parameter for diffusivity. Constrictivity was defined as a function of thickness of the montmorillonite layer based on the experiments of Sato et al. (1995). The thickness can be obtained from the dry density and properties of bentonite based on the equation proposed by Komine and Ogata (1999).

Figure 7-25 shows analytical results that investigate the effect of the surrounding bentonite on the long-term calcium leaching from concrete (Nakarai et al. (2003)). The results imply the possibility of accelerated leaching by the surrounding bentonite. This is because a low concentra-

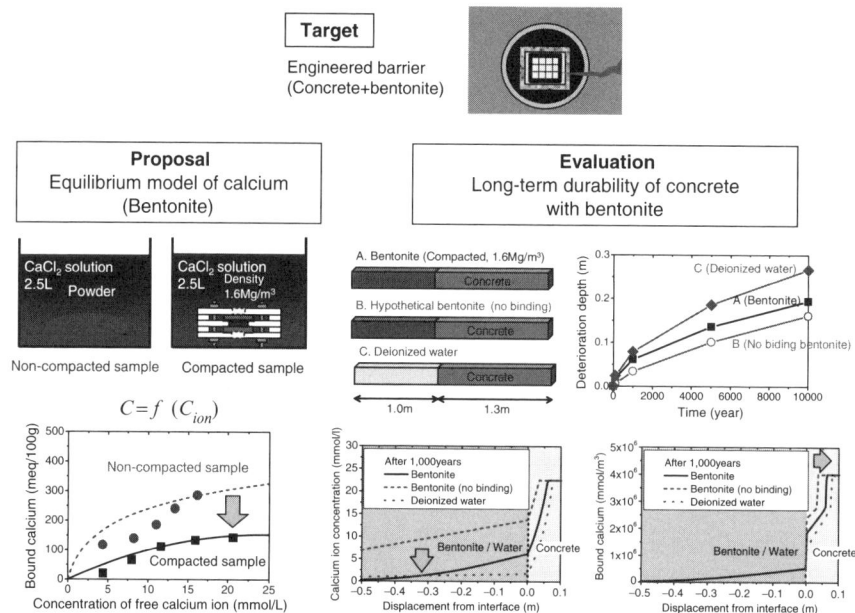

Fig. 7-25. Effect of bentonite on leaching of concrete

tion of free calcium ions in the pores of the surrounding bentonite was maintained due to the high binding capacity of ions and the constantly high concentration gradient between the concrete and the bentonite. Although the bentonite may protect concrete against any water penetration from nature, it simultaneously weakens the surface layer of concrete. Then, the multiphase modeling that can treat both main physicochemical events of mother materials and their interaction plays a substantial role for life-cycle assessment in terms of the composite barrier system.

7.6 Conclusions

Physicochemical and mechanical modeling of concrete with greatly different scales of geometry was presented, and synthesized on a unified computational platform, which may facilitate a quantitative assessment of structural concrete performances of interaction with soil foundation. The safety assessment method was extended to the life-cycle issue with multiscaled information about microclimate states of cementitious composites under macroscopic ambient boundary conditions. Currently granted is a great deal of knowledge earned by past development. At the same time, we face a difficulty to quantitatively extract consequential figures from them. The author expects that the systematic framework on this knowledge-based technology will be extended efficiently and can be steadily taken over by engineers in charge.

Acknowledgements

The author's appreciation is extended to Dr. Kishi for valuable discussion. This study was financially supported by Grant-in-Aid for Scientific Research (S) No.15106008.

References

Collins, M.P. and Vecchio, F. (1982) *The Response of Reinforced Concrete to In-Plane Shear and Normal Stresses*. Toronto: University of Toronto Press

El-Kashif, K.F. and Maekawa, K. (2004) "Time-dependent post-peak softening of RC members in flexure", *Journal of Advanced Concrete Technology*, 2(3): 301–315.

Guo, X.H. and Gilbert, R.I. (2002). An experimental study of reinforced concrete flat slabs under sustained service loads. UNICIV Report No. R-407

Japan Society of Civil Engineers (JSCE) (1999) *Recommendation for Structural Performance Verification of LNG Underground Storage Tanks*. Tokyo: Concrete Library of JSCE, No. 98

Japan Society of Civil Engineers (JSCE) (2005a) *Report on Safety and Serviceability of ASR Damaged RC Structures*. Tokyo: Concrete library of JSCE, No.124

Japan Society of Civil Engineers (JSCE) (2005b) Intermediate technical report on damage and repair of Tarui highway bridge. Special subcommittee report

Kakuta, Y. and Fujita, Y. (1982) "Fatigue strength of reinforced concrete slabs failing by punching shear", *Proceedings of JSCE*, (317): 149–157

Komine, H. and Ogata, N. (1999) "Experimental study of swelling characteristic of sand-bentonite mixture for nuclear waste disposal", *Soils and Foundations* 39(2): 83–97

Mabrouk, R., Ishida, T. and Maekawa, K. (2004) "A unified solidification model of hardening concrete composite for predicting the young age behavior of concrete", *Cement and Concrete Composites*, (26): 453–461

Maeda, Y. and Matsui, S. (1984) "Fatigue of reinforced concrete slabs under trucking wheel load", *Proceedings of JCI* (6): 221–224

Maekawa, K., Pimanmas, A. and Okamura, H. (2003) *Nonlinear Mechanics of Reinforced Concrete*. London: Spon Press

Maekawa, K., Chaube, R.P. and Kishi, T. (1999) *Modeling of Concrete Performance*. London: Spon Press

Matsui, S. (1987) "Fatigue strength of RC-slabs of highway bridge by wheel running machine and influence of water on fatigue", *Proceedings of JCI*, 9(2): 627–632

Nakarai, K., Ishida, T., Maekawa, K., et al. (2005a) "Calcium leaching modeling of strong coherency with micropore formation of porous media and ion phase equilibrium", *Journal of Materials, Concrete Structures and Pavements, JSCE*, 802(69): 61–78

Nakarai, K., Ishida, T. and Maekawa, K. (2005b) "Multi-phase physicochemical modeling of soil-cementitious material interaction", *Journal of Materials, Concrete Structures and Pavements, JSCE*, 802(69): 137–154

Nakarai K, Usui T, Ishida T (2003) Performance evaluation of engineered barrier based on thermodynamic modeling of coupled cementitious material and bentonite. The Tenth East Asia-Pacific Conference on Structural Engineering and Construction (EASEC-10), 2003.8

Perdikaris, P.C. and Beim, S.R. (1988) "RC bridge decks under pulsating and moving load", *Journal of Structural Engineering ASCE*, 114(3): 591–607

Perdikaris, P.C., Beim, S.R. and Bousias, S.N. (1989) "Slab continuity effect on ultimate and fatigue strength of reinforced concrete bridge deck models", *Structural Journal ACI*, 86(4): 483–491

Pimanmas, A. and Maekawa, K. (2001) "Finite element analysis and behavior of pre-cracked reinforced concrete members in shear", *Magazine of Concrete Research*, 53(4): 263–282

Sato, H. et al. (1995) "Diffusion behavior for Se and Zr in sodium-bentonite", *Scientific Basis for Nuclear Waste Management*, XVIII(353): 269–276

Satoh, Y. et al. (2003) "Shear behavior of RC member with corroded shear and longitudinal reinforcing steels", *Proceedings of the JCI*, 25(1): 821–826

Soraoka, H., Adachi, M., Honda, K., et al. (2001) "Experimental study on deformation performance of underground box culvert", *Proceedings of the JCI*, 23(3): 1123–1128

Sugano, T. et al. (2001) "Numerical analysis on uneven settlement and seismic performance of Tokyo brick bridges", *Structural Engineering Design* (17) JR-East: 96–109

Toongoenthong, K. and Maekawa, K. (2005a) "Multi-mechanical approach to structural performance assessment of corroded RC members in shear", *Journal of Advanced Concrete Technology*, 3(1): 107–122

Toongoenthong, K. and Maekawa, K. (2005b) "Simulation of coupled corrosive product formation, migration into crack and its propagation in reinforced concrete sections", *Journal of Advanced Concrete Technology*, 3(2): 253–265

Toongoenthong, K. and Maekawa, K. (2004) "Interaction of pre-induced damages along main reinforcement and diagonal shear in RC members", *Journal of Advanced Concrete Technology*, 2(3): 431–443

8. Earthquake-Resistant Engineering of Steel Structures

Hitoshi Kuwamura

8.1 Introduction

Steel structures are predominant in building construction in Japan. As shown in Fig. 8-1, steel structures increased rapidly after World War II and at present account for about 40% of the total floor areas of newly constructed buildings in a year. This ratio is comparable with wooden structures, while concrete structures are limited to about 20%. Such a high share of steel in building construction is unique in the world. The reason why steel buildings are so popular in Japan may be found in the strong support the government gave to developing the steel industry soon after the Meiji Restoration. The policy invited an extraordinary advancement of the technology of applying steel materials to various industrial products. At the same time, however, another important factor cannot be neglected: Japan is an earthquake-prone country located in one of the most quake-hazardous regions on Earth. Thus, the basic idea of structural design has been established: that strong and tough materials should be used in building structures to resist the impact of severe earthquakes. Obviously, steel is the best material to fit such requirements. Although steel is more expensive than concrete and timber, the practice of employing it in buildings has been sustained by the national wealth of Japan.

The origin of the basic idea of earthquake resistance is found in the construction of Mitsui Honkan (Fig. 8-2), which was built on the basis of lessons derived from the 1923 Kanto Earthquake. This building was designed such that earthquake forces are totally sustained by steel frames and post-earthquake fire is fully insulated by brick walls. The phrase"Buildings must be absolutely earthquake-resistant and fire-resistant"described the design concept at that time, one that has been inherited by today's structural engineers (Kuwamura (2002)).

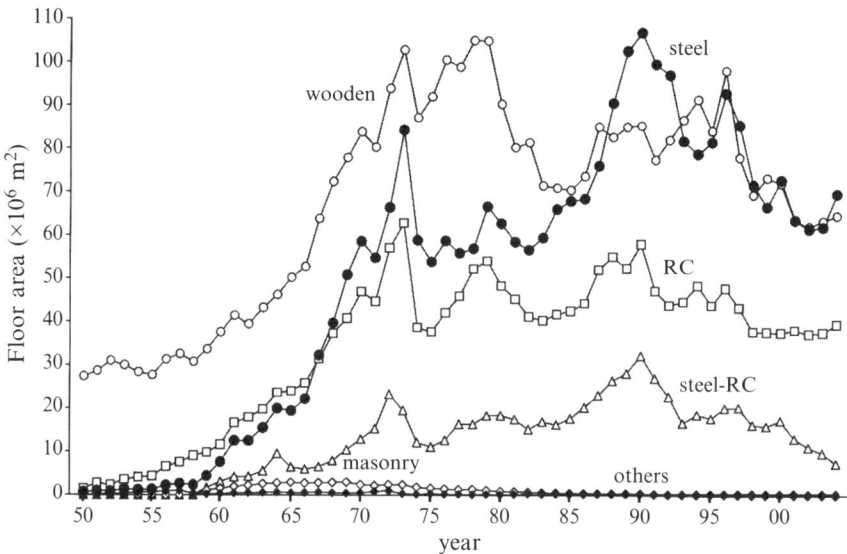

Fig. 8-1. Building construction in the latter half of the 20th century of Japan

Fig. 8-2. Mitsui-Honkan (Picture at 1929 completion, heritage of 1998 assignment. With permission from MITSUI FUDOSAN CO., LTD.)

Even in the field of bridge construction, steel piers are often seen supporting urban highways on short-span girders, while reinforced concrete piers are common practices in advanced foreign countries. This obviously comes from the earthquake-resistant concepts prevalent in Japan (Usami (2006)).

8.2 Damage to Steel Structures Caused by Earthquakes

Earthquakes are still serious natural hazards of unpredictable intensity that defy human understanding. Even steel structures cannot be totally free from seismic risk. The largest scale of damage in steel structures was experienced during the 1995 Hyogoken-Nambu Earthquake (Architectural Institute of Japan (1997)). This earthquake was a devastating near-field earthquake that devastated a big city of Japan, Kobe, that had been enjoying continuous prosperity after World War II. Many buildings as well as infrastructures were seriously damaged by the earthquake, and building collapses, mostly of wooden houses, killed more than 6,000 citizens. The damage to steel building structures can be categorized into two groups: one whose causes and outcomes can be explained from current knowledge, and another that cannot.

Introduced here are three buildings, the causes of whose failure can be explained (Kuwamura (2002); (Kuwamura Lab (1995)). Figure 8-3(a) shows the collapse of a four-storey residential building of steel construction. The first and second stories have been totally crushed without any original shape. The double faults in the bracings and the columns, as sketched in Fig. 8-3(b), account for the collapse. The faults in terms of connection fracture of the bracing and column are evidenced in Fig. 8-3(c) and Fig. 8-3(d), respectively. The cause of the brace-end fracture is that fracture strength Pu sustained by the net sectional area of the bolted connection is smaller than the yield strength Py of the gross sectional area of the bracing. Thus, the brace could not provide a sufficient amount of plastic elongation to resist the seismic attack. If the condition of seismic connection is satisfied, i.e., $Pu > Py$, such a fatal failure can be prevented. The fracture of the column base is due to the insufficient size of the fillet weld, since the weld deposit is too small to transfer the column stress to the foundation. If the fillet size is large enough, or a full-penetration weld is deposited, such a crucial failure can be avoided.

Figure 8-4(a) shows an overturned four-storey office building. The failure process as shown in Fig. 8-4(b) was triggered by the fracture of the welded connection of the column as evidenced in Fig. 8-4(c). The connection was too weak to sustain the earthquake force introduced against the column. At present, it is common sense to employ full-penetration weld to such an important connection of a main frame.

The third example as shown in Figure 8-5(a) is a parking-lot building seriously distorted due to the fracture of bracing-to-column connections. This is a typical example of poor detailing of steel connections. As shown in Figure 8-5(b), the gusset plate joining the bracing and the column was

Fig. 8-3. Collapse of a four-story steel residential building

welded to the web plate of the column, and consequently the web plate was pulled out in the perpendicular direction of its plane and ruptured. An adequate stiffening of the column web was necessary so that the tensile force in the bracing is smoothly transferred into the column.

The cause of failure due to brittle fracture is not well understood given the current state of knowledge. Figure 8-6(a) shows high-rise residential buildings constructed in a mega-structural system with the use of truss-type steel columns. The main columns underwent through-section cracking by the earthquake as shown in Fig. 8-6(b) and Fig. 8-6(c). Figure 8-7(a) shows the same mode of fracture in the lower flange of a beam in an ordinary moment frame for office use. This type of fracture was observed in more than 100 steel building frames in and around Kobe City and was a controversial issue among structural engineers in those days. Experts guessed the cause differently such as the lack of material toughness, cold temperature, strain concentration at the vicinity of welded joints, poor welding workmanship, large building drift caused by the epicentral earthquake, and large strain rate due to vertical shock waves. A scientist picked up a sample from the damaged building, shown in Fig. 8-7(b),, and observed the fracture surface with a magnification of 1,000 by means of a scanning election microscope.

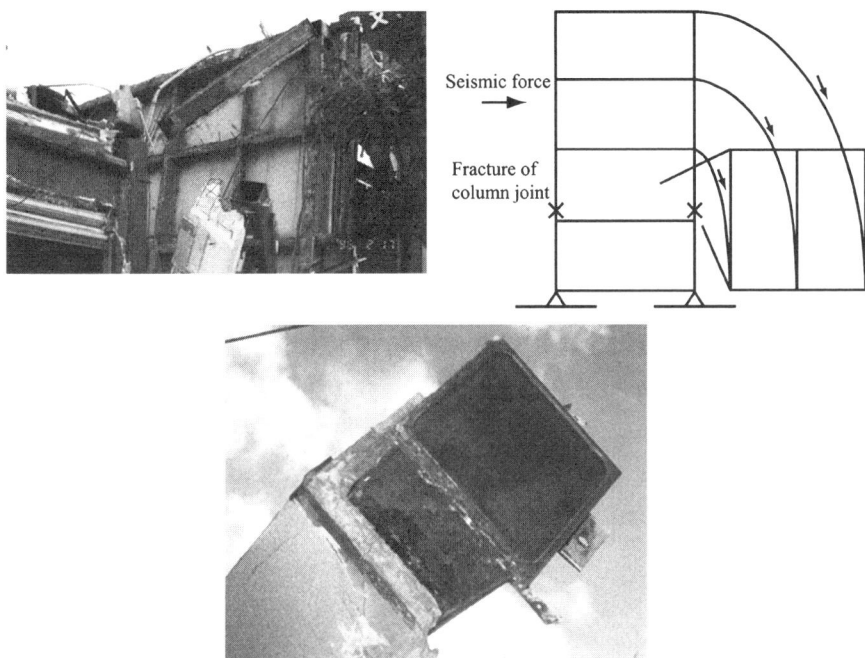

Fig. 8-4. Overturn of a four-storey steel office building

Fig. 8-5. Distortion of a parking lot building

Fig. 8-6. Brittle fracture of columns in a mega-structure

He identified a dimple pattern of Figure 8-7(c) at the origin of fracture and a river pattern of Fig. 8-7(d) over the area of fracture propagation. The same patterns of brittle fracture were found in a steel specimen he tested in his laboratory. From this observation, it was clarified that the brittle fracture is triggered by a micro-ductile crack generated during seismic vibration. A constitutional equation for crack initiation has been calculated from stress-strain analysis, while the condition of transition from ductile cracking to brittle fracture has not yet been quantified. At present, a tentative provision is proposed to prevent brittle fracture, one in which material, welding, and detailing are ranked from poor to excellent quality, respectively (Architectural Institute of Japan (2000)); (Kuwamura (2003)). It was fortunate that none

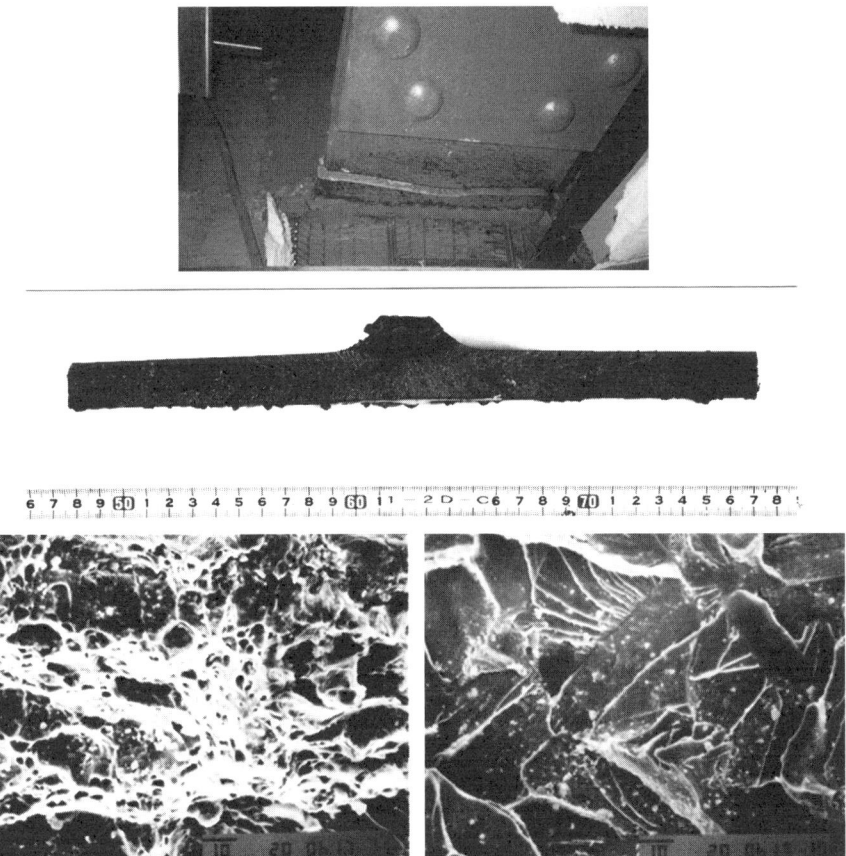

Fig. 8-7. Brittle fracture of a beam in a moment frame

of the steel building frames suffering from such brittle fracture collapsed. However, the reasons for this have not been clarified. The fail-safe mechanism that fracture of some structure elements does not lead to system failure is an important issue in seismic design.

In the field of bridge construction, the same type of damage also occurred during the 1995 Kobe Earthquake. Figure 8-8 shows the buckling of plat plates in a pier of rectangular hollow section, and Figure 8-9 shows the so-called elephant foot bucking of a pier of circular hollow section. The distortion took place in the thinner plates at the place where the thickness alternates. Since steel bridge structures are of thin-walled construction, they are most susceptible to plate instability during the overload of earthquake. Figure 8-10 shows the fatal collapse of a box-section pier due to progressive tearing of corner welds. Figure 8-11(a) and Fig. 8-11(b) show the brittle fracture of columns of centrifugally cast steel pipe that supported a railroad.

Fig. 8-8. Buckling of steel pier of rectangular hollow section

8.3 Fundamentals of Steel Structural Safety

The safety of steel structural buildings against external disturbances, especially earthquakes, is governed by the total coordination of three factors: adequate selection of steel material, quality control of fabrication (especially welding), and technology of structural design. Japan, the most earthquake-prone country among highly industrialized nations, has advanced the holistic research and development for the above three phrases, which are described in the following sections.

8.3.1 Steel Material Selection

Most of the structural steels employed in building construction are specified in JIS (Japan Industrial Standards). A new steel named"Rolled Steel for Building Structure"entered JIS in 1994. This steel was designed with

Fig. 8-9. Buckling of steel pier of circular hollow section

Fig. 8-10. Collapse of steel pier of rectangular hollow section

Fig. 8-11. Buckling of steel pier of centrifugally cast steel pipe

unique properties for earthquake resistance. This steel, as shown in Table 8-1, is designated SN followed by a number such as 400 and 490 that indicates tensile strength. The final symbols A, B, and C designate more detailed performance grades. The specific characteristics in the steel are seen in grades B and C, in that the yield ratios are limited to 80% and the yield points have upper as well as lower bounds. These properties are known to be effective for steel structures to resist a strong earthquake in terms of energy absorption after yielding.

A lower yield ratio means a larger extra-strength capability after yielding. Thus, a steel member made of a lower yield ratio exhibits a larger plastic deformation during the post-yielding strength, as shown in Figure 8-12. Thus, the upper limit of yield ratio (YR) is essential to develop member ductility.

In a member made of the steel with an upper bound to its yield point, a plastic hinge is certainly formed at the prescribed level of bending moment. This means that the ultimate mechanism of the designer's intention is realized in his structure and then such intended ductility is surely provided to the whole structural system. This is concretely demonstrated in Figure 8-13. The plots in Fig. 8-13(a) show that the yield points of conventional steels are scattered in a considerably larger range for smaller thickness. However, when the range of dispersion becomes narrower like Range C (COV = 10%) to Range A (COV = 2.5%) of Fig. 8-13(a), the system ductility is significantly improved as shown in Fig. 8-13(b) and Fig. 8-13(c) (Kuwamura and Sasaki (1990)). In Fig. 8-13(b), three types of ultimate behavior are shown. Types 1, 2, and 3 have horizontal sway of one story, three stories, and the whole six stories, respectively, under earthquake forces. Type 3 exhibits the highest strength in terms of load factor, but also the largest deformation in terms of top displacement. This suggests that system ductility is increased with the number of plastic sway stories involved in the ultimate behavior. The result of statistical simulation of Fig. 8-13(c) shows that the reduction

Table 8-1. JIS specification of "Rolled Steels for Building Structures (SN steel)"

Symbol	Yield point (N/mm²) Plate thickness, t (mm) 6≤t<12	12≤t<16	16<t≤40	40<t≤100	Tensile strength (N/mm²)	Yield ratio (%) Plate thickness, t (mm) 6≤t<12	12≤t<16	16<t≤40	40<t≤100	Elongation (%) No.1A 6≤t≤16	No.1A 16<t≤50	No.1A No.4 16<t≤40	No.1A No.4 40<t≤100	Reduction of area (%) Plate thickness, t (mm) 16≤t≤100	Charpy impact energy (J) Specimen no.4 rolling direction
SN400A		235 or greater	215 or greater		400–510		No limit			17 or greater	21 or greater	23 or greater	23 or greater	No limit	No limit
SN400B	235 or greater	235–355	235–355	215–335		No limit	80 or less			18 or greater	22 or greater	24 or greater	24 or greater		27 J or greater
SN400C	—	235–355	235–355			—		80 or less						Average of three 25 or greater (each 15 or greater)	
SN490B	325 or greater	325–445	325–445	295–415	490–610	No limit	80 or less			17 or greater	21 or greater	23 or greater	23 or greater	No limit	
SN490C	—	325–445	325–445			—		80 or less						Average of three 25 or greater (each 15 or greater)	

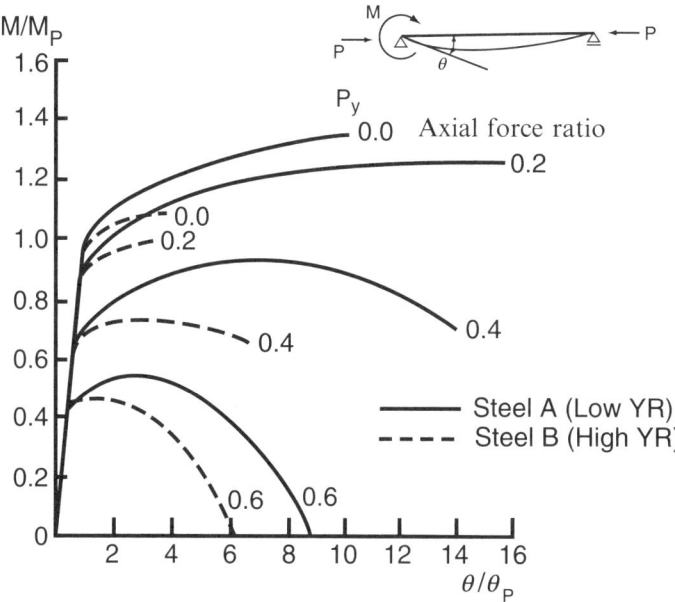

Fig. 8-12. Yield ratio vs plastic deformation capacity of steel member

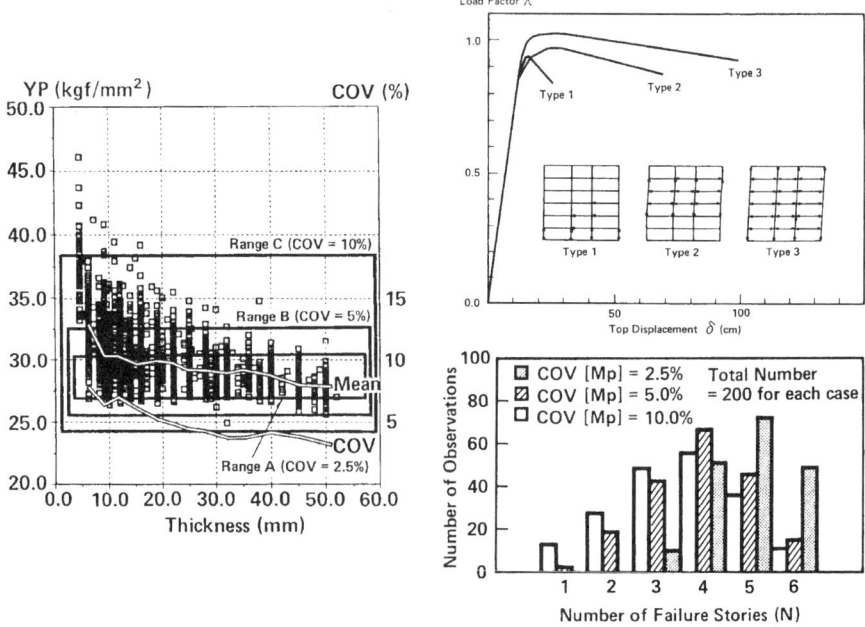

Fig. 8-13. Range of yield point vs Plastic deformation capacity of steel frame

of yield point dispersion in terms of COV (coefficient of variation) brings the statistical shift of larger number of sway stories involved in the ultimate behavior. Thus, the upper limit of yield point (YP) is very important for developing system ductility. The earthquake- resistant design on the basis of weak beam concept, which is mentioned in the latter discussion, should be associated with the use of the YP-control steel.

8.3.2 Welding

Needless to say, strength is fundamental for welded joints. In addition to strength, alterations of material due to welding heat input as well as configurations of welded joints are taken care of in the steel construction of medium- to high-rise buildings in Japan. Such thoughtful quality control may not be seen in other countries, because Japanese bitter experience from past earthquakes impelled the careful fabrication.

Figure 8-14 shows the relationships between welding heat input and toughness in terms of Charpy impact energy at heat-affected zones (Kuwamura (2002)). The welding heat input tends to increase in this order: manual welding, CO_2 gas shield arc welding, submerged arc welding, electro-gas welding, and electro-slag welding; from that point the toughness tends to decrease in the same order, since the welded connections from larger heat input are more susceptible to brittle fracture. Welding with a large heat input enhances the productivity of steel fabrication, but it may cause problems in seismic performance of steel structures. Thus, the amount of heat input must be limited and time-consuming workmanship practiced with care about material properties.

Figure 8-14 shows various configurations at beam-to-column welded connections subjected to severe stresses during earthquake motion. The detailings in Fig. 8-15(a) are the conventional ones, but now it is known that these connections are susceptible to fracture, because sharp notches at the toes of fillet weld can become a cause of brittle fracture. Thus, the improved connections in Fig. 8-15(b) are now in place.

8.3.3 Design

The safety of steel structures against a big earthquake is largely dependent on the sophistication of structural design. Structural control, or the technique of absorbing earthquake input energy is decisive. Recently, structural designers have prescribed the ultimate behavior of their buildings and identified the

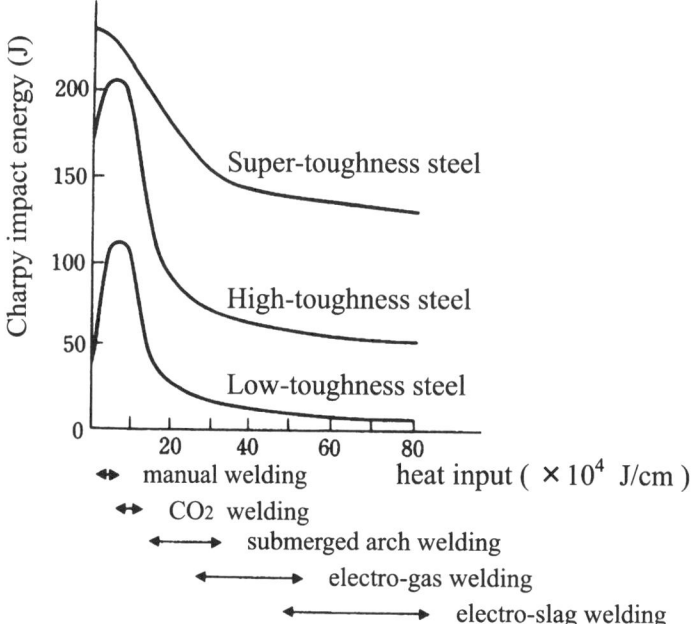

Fig. 8-14. Welding heat input vs fracture toughness

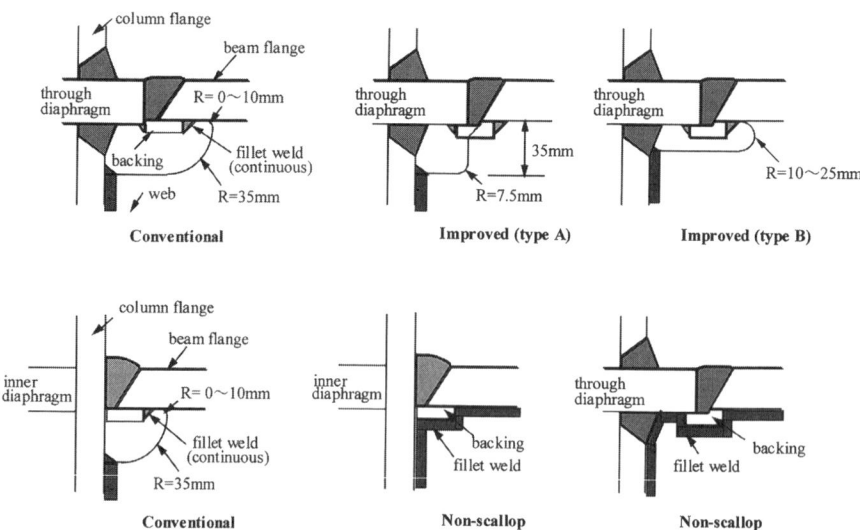

Fig. 8-15. Improvement of connection detailing

structural members forced into plastic strain. Such members are designed to have enough ductility to dissipate earthquake energy. Figure 8-16 shows an example of such structural control, which is called a weak-beam structure. This type of seismic structure is practiced in reinforced concrete as well as in steel construction. The columns are proportioned to be stronger than the beams. When the building is attacked by a severe earthquake, plastic hinges are formed at beam ends and horizontal drift is sustained by the cooperation of all stories. This means that the total building system exhibits a very tough resistance. If the columns are weak, a local inter-story failure may immediately occur. Actually, several reinforced concrete frames collapsed in the mode of local inter-story failure due to weak columns during the 1995 Earthquake.

Recently, base-isolated structures as schematically shown in Figure 8-17(a) have become popular. However, it costs more than ordinary construction due to double layer foundations for installing the isolation, and the resonance to long-period seismic wave is apprehended. In order to avoid the double foundations, a new type of isolation as shown in Fig. 8-17(b) has been invented, in that the first story is provided for isolation. Figure 8-17(c) shows a typical vibration-controlled system, one that uses dampers for energy dissipation in steel bracings. Such advanced technology in seismic design is indebted to newly developed damper materials and also computer-aided earthquake response analysis.

8.4 Future Seismic Design

The concept of seismic design of buildings constructed of steel as well as other materials is now in a revolutionary transition from purposes of life-saving only to one that also includes property reservation, which is called

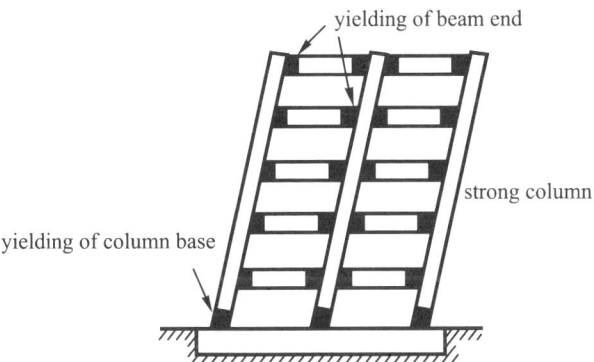

Fig. 8-16. Earthquake-resistant structure of weak-beam concept

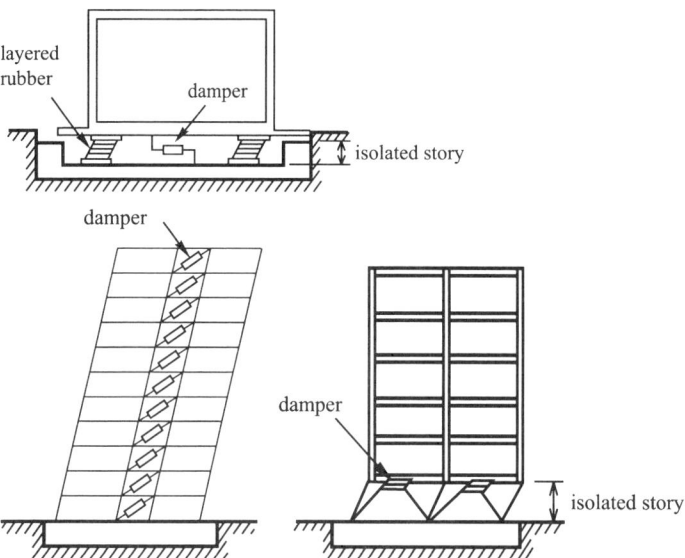

Fig. 8-17. Base-isolated structure and vibration-control structure

Performance-Based Design. The original idea for this concept is found in the activities of the Vision 2000 Committee organized in 1992 by the Structural Engineers Association of California (SEAOC), based on the lessons from the 1989 Loma Prieta earthquake that attacked the coastal area around San Francisco. Since this earthquake was not so severe (magnitude 7.1), the safety of buildings designed to meet prevailing codes was assured, though it caused economic losses of US$7 billion. The study of performance-based design originated from the necessity of including the criterion of property preservation in addition to life safety in the framework of seismic design. The Committee submitted a final recommendation in 1995, in which holistic seismic engineering of construction and maintenance as well as design is schemed (Structural Engineers Association of California (1995)).

The 1994 Northridge Earthquake (M6.7) caused a larger economic loss (US$20 billion) than the Loma Prieta Earthquake and enhanced the activities of the Structural Engineers Association of California (SEAOC). In addition, the earthquake revealed the brittle fracture problems of steel building frames. It had been believed that steel moment frames were ductile enough to resist a big earthquake, but a lot of beam-to-column connections underwent brittle fracture causing an unacceptable level of losses of building properties despite little loss of human life. In order to establish a provision for preventing brittle fracture in steel moment frames, the SAC Joint Venture of SEAOC, ATC, and CUREe was organized in 1994 on a research fund from the Federal Emergency Management Administration (FEMA). The SAC Joint Venture submitted a final report in 2000, in which a framework of performance-based design was proposed with an emphasis on seismic performance of steel moment frames (SAC Joint Venture (2000)).

In Japan, in accordance with the American movement of performance-based design, a national project named "Development of New Building Structural Systems" was started in 1995 by the initiative of Ministry of Construction, and the study on performance-based design was promoted for three years. The necessity of performance-based design in Japan was compelled by two events: the 1995 Hyogo-ken Nambu Earthquake (M7.2) which caused huge economic losses of more than ¥10-Cho (about US$100 billion) as well as more than 6,000 human deaths; and the 1996 US-Japan top-level conference about the business of building construction being open to the free market. The final report of the national project was published in 2000 (Ministry of Construction (2000)). The basic concept in the report is similar to that of the Vision 2000 Committee, and is reflected in the amendment of the Building Standard Law issued in 1998 and enforced in 2000. Building design is now moving from being specification-based to being performance-based.

As mentioned above, the aim of performance-based design is to preserve building properties together with human lives. However, the essence of performance-based design lies in mutual recognition between designers and customers about building performance. In order to preserve building properties against earthquake, an adequate extra amount of money should be invested to the construction. Thus the designers should be responsible for the accountability on any cost vs. effect issues. In such a situation, the interface between designer and customer is the performance matrix. Several forms of the performance matrix were proposed by SEAOC Vision 2000, SAC-FEMA, and the Japan National Project, and are not yet unified, but the concept is a shared one, as shown in Table 8-2 (Kuwamura et al. (2002)). The row (horizontal axis) in the matrix indicates the level of external load in terms of the probability of occurrence, and the line (vertical axis) indicates the level of performance in terms of the degree of damage. The target performance of the building structure of concern is determined by placing marks in the matrix, like single-hollow circles. Since a larger intensity of lead obviously causes heavier damage, a set of marks is placed downward to the right-hand side. When the set of single-hollow circles represents a standard level of performance, the sets of double-hollow circles and star marks indicate higher levels of performance.

Table 8-2. Performance matrix

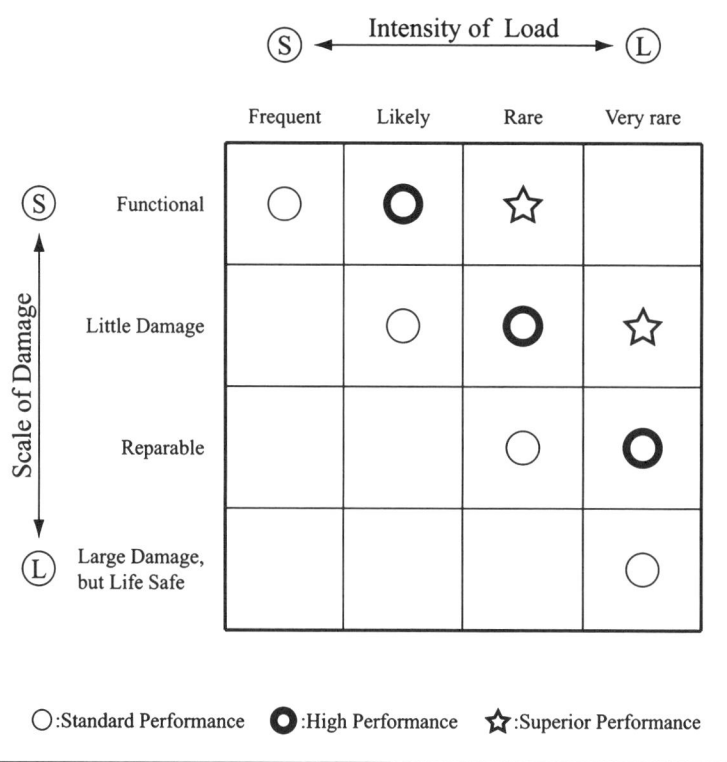

O:Standard Performance   ●:High Performance   ☆:Superior Performance

The performance level, which constitutes an axis of the performance matrix, is correspondent to the degree of damage resulting from the application of external loads. Thus, the performance level ranges widely from healthy state to fatal collapse. The lower bound of a performance level, i.e., the limit beyond which the performance cannot be kept, is called the limit state. The definition of performance levels is now subject to various proposals, but the following three are considered to be essential: (1) the classified performance should be meaningful to the customers, (2) the associated damage should be identifiable, and (3) the limit state should be quantifiable by means of engineering mathematics. The performance levels shown in Table 8-3 are proposed for steel structural buildings to meet the above-mentioned three requirements (Kuwamura et al. (2002)). The performance is ranked into four levels: Level 1, Functional; Level 2, Undamaged; Level 3, Reparable; and Level 4, Life Safe. Level 5, Collapse is not permissible except for special cases such as warehouses without human accommodation.

The limit states associated with Performance Levels 1–4 are called function limit, damage limit, repair limit, and safety limit, respectively.

From the viewpoint of stock preservation, the performance of reparability has recently been noticed as important. If a building damaged by an earthquake can be repaired and reused without demolishing, the economic loss will be significantly reduced and furthermore environmental impact due to constructional wastes will be diminished. The reparability design, which has not been considered in the conventional design philosophy, first appeared in the framework of performance-based design. From the analysis of damage and repair costs to steel-framed buildings that suffered moderate damage from the 1995 Kobe Earthquake, the reparability limit state was quantified. The reparability for reuse is found to be judged from the residual drift after an earthquake. When overall residual drift and maximum inter-story residual drift are not more than 1/200 and 1/90, respectively, the building can be repaired with acceptable cost and current technology (Iwata et al. (2005)). The problem that remains is to establish a method of predicting the residual drift as exactly as possible in the design procedure.

8.5 Conclusive Remarks with Seismic Rehabilitation of Existing Structures

There exist a lot of inadequate structures in terms of earthquake resistance. They were constructed on the old design code of low demands and have possibly deteriorated during long service periods. Many types of rehabilitation techniques for old steel structures are going to be developed, such as strengthening with supplemental structural elements and adding of base isolation or energy absorbing devices.

A part of the 2004 supplementary government budget was assigned to earthquake rehabilitation of Building No.11 of the School of Engineering at the University of Tokyo (Fig. 8-18). Construction for strengthening was started in May 2005 and was completed at the end of December 2005. In earthquake-prone countries like Japan, the technology of seismic retrofitting of existing buildings is substantial for the sustainability of urban society. The current advanced seismic technology of Japan was applied to the retrofitting of Building No.11, from which a model of seismic rehabilitation was proposed in the COE program Creation of New Society and Engineering for Sustainable Urban Regeneration (Department of Architecture, School of Engineering, the University of Tokyo (2005)).

Table 8-3. Performance description for steel buildings

Performance level		Building as a whole	Structural skeleton	Damage state of building parts			
				Non-structural elements	Facilities	Furniture	Remarks
Level 1	Functional	Continuous service is retained without any discomfort and disturbance.	No damage occurs in structure. Deflection and vibration are minor to keep comfort.	No damage.	Operational as usual.	Perfectly protected.	No damage and no inconvenience for users.
(Function limit)							
Level 2	Undamaged	Slight damage is observed, but basic function is sustained for normal service.	Negligible damage such as micro yielding at stress concentration may takes place. Deformation is within elastic limit with no residual deflection. Inspection and repair are unnecessary.	Little crack or minor peeling may occur, but any malfunction such as leak of rain water does not occur. Repair may be done at users convenience.	Emergency lock may work, but immediately restart. No malfunction of safety equipment such as sprinklers.	Slide and overturning of light furniture may occur, but no major damage. Hazardous materials are protected.	Temporary inconvenience and little economical losses may occur. Normal services can be resumed immediately (green tag).
(Damage limit)							

Level 3	Repairable	Apparent damage is observed, but is technically and economically reparable. Original performance is recovered and building is reusable as before. Minor injury may occur.	Elastic limit is exceeded, and some extent of yielding, buckling and fracture are partly observed. The structural capacity is moderately impaired.	Moderate to severe damage occurs, but are reparable or replaceable. Large falling hazards are not created.	Many facilities are damaged and non-operational, but can be repaired by experts.	Many of fixtures are damaged due to slide and overturning, but valuable or hazardous materials are protected.	Continuous occupation after damage may be inadequate, which is judged only by experts (yellow tag).
(Repair limit)						
Level 4	Life safe	Major damage occurs, but building sustains its own weight. Floor and roof do not fail. Human lives are protected, while some are wounded.	Extensive yielding, buckling and fracture occur, but the frame can withstand gravity load pertinent to building.	Extensive damage occurs, but fatal falling hazards are not created. Egress may be obstructed, but self-escape and rescue activities are possible.	Emergency facilities for human lives are operational, but others are non-operational.	Many are seriously damaged, but dangerous materials do not spill.	Off limits (red tag).
(Safety limit)						

(continued)

Table 8-3. (continued)

	Damage state of building parts					
Performance level	**Building as a whole**	**Structural skeleton**	**Non-structural elements**	**Facilities**	**Furniture**	**Remarks**
Level 5 Collapse	Major parts or whole of building collapse, and human lives are exposed to critical state.	Serious damage causes major loss of structural capacity for sustaining building own weight.	Falling hazards are fatal. Egress may be seriously obstructed.	Almost all facilities are fatally damaged and non-operational.	Almost all are substantially damaged. Hazardous materials may spill out.	This state is not socially permitted except for special buildings in which no humans and no hazardous materials are contained.

Fig. 8-18. Building No.11 of school of engineering (before seismic rehabilitation)

References

Architectural Institute of Japan (1997) Report on the Hanshin-Awaji Earthquake Disaster, Building Series Vol 3, Structural Damage to Steel Buildings, Maruzen

Architectural Institute of Japan (2000) Tentative Provisions for Brittle Fracture of Steel Buildings, Proceedings of Panel Discussion: 9.

Department of Architecture, School of Engineering, the University of Tokyo (2005) Report on Seismic Retrofitting of Building No 11 of School of Engineering: 3

Iwata, Y., Sugimoto, H. and Kuwamura, H. (2005) Reparability Limit of Steel Structural Buildings. Journal of Structural and Construction Engineering (Transactions of AIJ) 588(2): 165–172

Kuwamura, H. (2002) Performance and Design of Steel Structures. 1.2 History of Iron and Steel Structures, 11.10 Welding and Earthquake Damage, Kyoritsu-Shuppan

Kuwamura, H. (2003) Classification of Material and Welding in Fracture Consideration of Seismic Steel Frames, Engineering Structures. Journal of Earthquake, Wind and Ocean Engineering 25(5): 547–563

Kuwamura, H. and Sasaki, M. (1990) Control of Random Yield-Strength for Mechanism-Based Seismic Design. Journal of Structural Engineering, ASCE 116(1): 98–110

Kuwamura Lab (1995) Field Survey Report on 1995 Hyogoken-Nambu Earthquake Disaster

Kuwamura, H., Tanaka, N., Sugimoto, H. and Kouno, T. (2002) Performance Description of Steel Structures. Journal of Structural and Construction Engineering 562(12): 175–182.

Ministry of Construction (2002) Concept of Performance-Based Design of Building Structures, Gihodo-Shuppan

SAC Joint Venture (2000) Recommended Seismic Design Criteria for New Steel Moment-Frame Buildings, FEMA-350, Chap 2 General Requirements, Chap 4 Performance Evaluation

Structural Engineers Association of California (SEAOC) (1995) Sacramento, CA: Vision 2000 Committee: Performance Based Seismic Engineering of Buildings, Sacramento, CA

Usami, B. (2006) Guideline for Seismic and Damage Control Design of Steel Bridges, Gihodo-Shuppan

9. Wide-Roof Buildings in Earthquakes

Ken'ichi Kawaguchi

9.1 Introduction

In most cases "seismic performance of building structures" means structural behavior of "tall" buildings. Very few articles tell about the seismic performance of "wide" buildings. In this article the real situation of the seismic performance of wide-roof buildings are indicated and discussed based on the author's experiences.

Wide-roof buildings are usually used for gymnasiums, event halls or exhibition halls where many people gather for certain activities. Of course they often used for factories or warehouses with a very small population inside. However when they are used for the purposes of public gatherings, the safety of the building is very important. In this article, wide-roof buildings used for gathering purposes are mainly considered.

This picture of Mt. Fuji, a snow-capped volcanic mountain, is a proof that Japan is in a real seismic region (Fig. 9-1). Japan has four distinct seasons, with hot and humid summers and cold winters with heavy snow, especially in the mountainous areas in the north. Typhoons are experienced every summer. The ground moves, wind blows, and water flows, exposing Japan to severe natural conditions, so its our structures must be prepared against most of the forces brought about by these conditions.

Before a discussion about earthquake damage, it should be pointed out that the structures with wide roofs are usually more severely damaged by snow or wind than by earthquakes. Wide roofs are usually designed as lightweight structures covering wide areas. Snow load and wind load usually impact the structure by their area while inertial forces during an earthquake impacts by their weight. For example in 1998 three school gymnasiums were collapsed by heavy snow in Yamanashi prefecture, as shown in Fig. 9-2. It was not in Japan but in Europe and Russia that several wide-roof buildings, such as swimming pool, ice arena and exhibition

Y. Fujino, T. Noguchi (eds.) *Stock Management for Sustainable Urban Regeneration*,

Fig. 9-1. A view of Mt. Fuji

Fig. 9-2. Wide-roof gymnasiums collapsed under heavy snow

hall, collapsed under heavy snow in 2006. Membrane roofs are often torn off and thin metal roof claddings are peeled off during typhoons. Hurricane Katrina in 2005 in the US may still evoke vivid images of these damages. On the other hand it is very rare that wide roof buildings collapse in earthquakes, although even such rare cases will be reported later in this article.

9.2 Seismic Designs of Wide-Roof Buildings

For seismic designs of wide roofs that spans larger than 100 m, a full dynamic response analysis is usually carried out during the design stage in order to check the structural safety under certain earthquake records, as specified by codes (see Table 9-1).

Table 9-2 shows examples of conditions that have been used for the seismic design of large-span structures wider than 100 m. The first natural periods calculated in the design stages—of about one second—are also indicated. This means they have some response during tremors since many response spectra of earthquakes have their peaks around one second. Therefore the seismic design of such structures must be properly carried out. The table also shows the criteria for seismic designs that are usually decided by structural designers by themselves under the guidance of authorities. Response analyses for typical real earthquake records are carried out, and their seismic performances are checked. The stress levels of structural components are usually kept under the permissible stress level of the material.

9.3 Damage to Wide-Roof Buildings by Some Earthquakes

9.3.1 Kobe Earthquake in 1997

After the Kobe Earthquake on 17 January 1995 we carried out a series of investigations of the damage to wide-roof buildings. The earthquake brought about devastating damage to many structures and killed more than 6,000 people. During the investigation we often saw heavily collapsed buildings that reminded us of the importance of proper seismic design, as shown in Fig. 9-3. The location of Kobe City is shown in Fig. 9-4.

9.3.1.1 Overview of Damage to General Structures

Many detached wooden houses have been severely damaged, as shown in Fig. 9-5. Since there had been an erroneous perception that Kobe was not in the seismic area. People in the Kobe area were rather less prepared against earthquakes and so were their houses. Old residential houses, shrines or temples with heavy ceramic roof claddings typically collapsed. Fire after the earthquake enlarged the damage to wooden structures considerably. It involved both collapsed and uncollapsed houses and caused calamities.

Table 9-1. Dynamic analyses for seismic design of wide-roof domes

		Fukuoka Dome Span 222 m	Kitakyshu Media Dome Span 142 m	Nagoya Dome Span 229 m
Modelling for calculation	Roof structure	Treated as a set of equivalent single-layer frames	Full frame and simplified frame	Full frame
	Supporting structure	Modeled as 12 sets of flexural and shearing springs	Full frame and simplified frame	Six-lumped-mass model with horizontal springs at base, considering foundations and ground stiffness
Seismic response analysis		Roof and supports are combined and calculated together	Roof and supports are combined and calculated at once	Roof and supports are separately calculated
Others		Horizontal foundation-ground interactions were considered	Either full frame or simplified frame is used	Mass of roof structure, which is 5% of the total mass, was considered as the mass of the top layer of the supporting structure

Table 9-2. Seismic design criteria for wide-roof buildings

		Fukuoka Dome	Kitakyushu Media Dome	Osaka Dome	Osaka Pool	Nagoya Dome	Komatsu Dome	Saitama Super Arena	Sapporo Dome
Site		Fukuoka	Kitakyushu	Osaka	Osaka	Nagoya	Komatsu	Yono	Sapporo
1st Natural Period (s)		1.032	0.705	0.96	0.439	0.568	0.951	1.48	1.14
Definitions	Level 1	160 gal	200 gal	20 cm/s	20 cm/s	25 cm/s	200 gal	200 gal	200 gal
	Level 2	320 gal	400 gal	40 cm/s		50 cm/s	400 gal	500 gal	400 gal
Seismogram used for dynamic response analyses	El Centro	○	○	○	○	○	○	○	○
	Taft	○	○	○	○	○	○	○	○
	Hachinohe	○	○				○	○	
	Others	Artificial seismogram		Osaka 205	Nagoya 306	ultimate load estimated by incremental analysis		Artificial seismogram	
Criteria	Level 1	allowable stress level for temporary load	allowable stress level	elastic region	elastic limit		elastic limit	elastic region	allowable stress level
	Level 2	allowable stress level with 1.1xF	with 1.1xF						with 1.1xF

Fig. 9-3. A completely collapsed reinforced concrete building

Fig. 9-4. Location of Kobe city

Many buildings with middle height were also severely damaged, shown in Fig. 9-6. Of special note was the observation that middle-height buildings experienced a distinctive failure, with the collapse concentrated at particular intermediate stories.

Fig. 9-5. Wooden houses damaged by Kobe earthquake

Fig. 9-6. Middle-story buildings damaged by Kobe earthquake

One of the most well-known damage involved the collapse of highways with cantilevered columns, as shown in Fig. 9-7. Besides these, other public structures, e.g., other kind of columns of highways, railways, quays at harbors and grounds on artificial landfills were also seriously damaged, as shown in Fig. 9-8. Industrial structures such as silos, as shown in Fig. 9-9, and facilities in plants, were also heavily damaged.

9.3.1.2 Structural Damage to Wide-Roof Buildings

In general it has been said that inertial forces during earthquakes do not affect wide-roof structures so much since they are designed as lightweight systems. Through our investigation we had the impression that this statement is generally justified. However there was some typical, but not so critical, damage that can be termed "structural," and there were also a few and very dangerous exceptions.

The wide-roof structures that we have been investigated were mostly built as steel-frame structures. These steel roofs are usually supported by either a reinforced-concrete skeleton or steel frames. Some of the vertical bays in these structures have supporting frames that are usually equipped with

Fig. 9-7. A highway supported by cantilever columns

Fig. 9-8. Public structures damaged by Kobe earthquake

seismic walls (for reinforced concrete skeletons) or braces (for steel frames). For reinforced concrete skeletons cracks in the seismic walls are commonly observed, and for steel skeletons, buckling and cutoff of these seismic braces. Roof frames are usually stiffened with horizontal braces in order to transmit horizontal in-plane forces due to wind and seismic loads. Buckling and cutoff of these horizontal braces are also commonly observed, as shown in Fig. 9-10. Since these bracing components are designed for additional loads of wind or earthquakes, they usually do not carry any self-weights. Therefore in most cases failure of these components never bring about any immediate collapse.

One of the weak points in the structures can be always found at "connections," especially where different materials are jointed together. For the wide-roof structures such connections frequently and unavoidably exist at bearings, where roof structures are buoyed by supporting structures. At bearings considerably large force must be transmitted from roof structures to the inferior structures. When the material of the roof structure, usually steel, is different from that of the supporting structure, often reinforced concrete, discontinuity of the deformation can easily occur especially under sudden change in the flow of the force. Fig. 9-11 shows some typical damage with the removal of covering concrete, observed at bearings where steel roof structures are supported by reinforced-concrete structures.

Fig. 9-9. Silos damaged by Kobe earthquake

In respect to structural stability this is not critical; however, any fall of concrete debris to the floor is very dangerous to the users of the buildings.

Three-dimensional truss systems called space frames are often employed for wide-roof structures. They are usually highly redundant systems with stress distribution that is not so simple as in usual regular structural frames. After the earthquake in Kobe, buckling and break of some members were observed in the space frames of roof structures, as shown in Fig. 9-12. Such damage was considered as minor since the frames still had enough number of members to support themselves and never reached failure mechanisms. On the other hand, stress distribution in the frames are rather complicated, and failed members were found at different locations in the frames—sometimes near bearings but sometimes at the midspan.

For the wide roof-structures built on artificial islands, as shown in Fig. 9-13, very little structural damage was observed, but ground sink that occurred at considerable depth caused problems for the architectural systems.

Fig. 9-10. Braces in steel frames

Fig. 9-11. Failure at bearings of supporting structure for roof frames

Fig. 9-12. Failed members of space frames

A pneumatic dome survived safely from the earthquake strike, as shown in Fig. 9-14. Of course they are very light, as light as balloons, so the inertial force induced by the earthquake had nothing to do with its structure. However, after the earthquake it suffered very much from the failure of the power supply. Pneumatic structures always need to be pressurized by blowers in order to keep the air pressure inside. The dome had an original self-powered generator and necessary stock of heavy oil for three days. But after the earthquake the power failure continued for one week. It became a very severe problem to keep internal air pressure by running the blowers. Lots of efforts were made to supply heavy oil to the dome since the dome was used as the relay station for relief supplies. This became a new type of problem observed for a pneumatic structure, which was thought to be very stable against an earthquake disaster.

The most dangerous failure observed for the wide-roof structure was the failure of precast roof systems. This school gymnasium in Fig. 9-15 had a precast roof consisting of 17 pieces of precast beams. This so-called Silberkhul system was imported from Germany in the 1960s. During the earthquake these precast pieces shifted separately and fell into the interior space. Fig. 9-16 shows the view of this precast roof and interior space after the earthquake. Anchors of the beams to the bearing walls exhibited poor strength, and the beams on the middle bays could easily slide off. The precast

Fig. 9-13. Wide-roof buildings on an artificial island

Fig. 9-14. A pneumatic dome

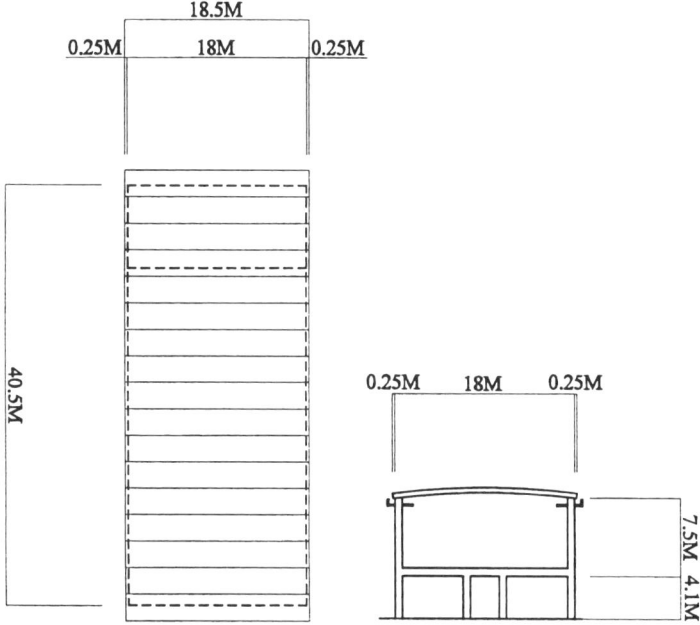

Fig. 9-15. School gymnasium with a precast beam roof

Fig. 9-16. View of the school gymnasium with a precast beam roof after the earthquake

beams were not connected to each other; therefore, they moved separately since they could not transmit the horizontal force to the end walls.
Fortunately, the earthquake occurred early in the morning and nobody was
inside of the gymnasium. However, it can easily be imagined how bad it
would have been if the earthquake had occurred during daytime when the
gymnasiums were in use.

The same precast beams were used at another public gymnasium in a
double-span construction using a keel girder crossing over the midspan, as
shown in Fig. 9-17. An interior view of the gymnasium after the earthquake
is shown in Fig. 9-18. One can see that most of the roof has fallen onto the
floor. Here again, connections of beams to the bearing walls and the keel
girder were so poor that they could not resist the inertial force that acted
during the earthquake against the rather heavy roof structures. One can
also observe the remains of the movement of walls during the earthquake
by tracing the crack pattern going through the walls and the columns. The
hinge line shows the bending deformation of the walls during the earthquake that shook off the roof beams. Anyone inside a gymnasium with such
a roof could hardly find space for escape during a heavy earthquake.

Fig. 9-19 shows an example of damage to another type of precast
beams used for school gymnasiums. This damage is much less than the
previous two cases; however, it still exhibits the traces of strong interaction between the beams and bearing girders. Precast reinforced-concrete
beams are much heavier than steel-frame roofs and they must be designed
more carefully.

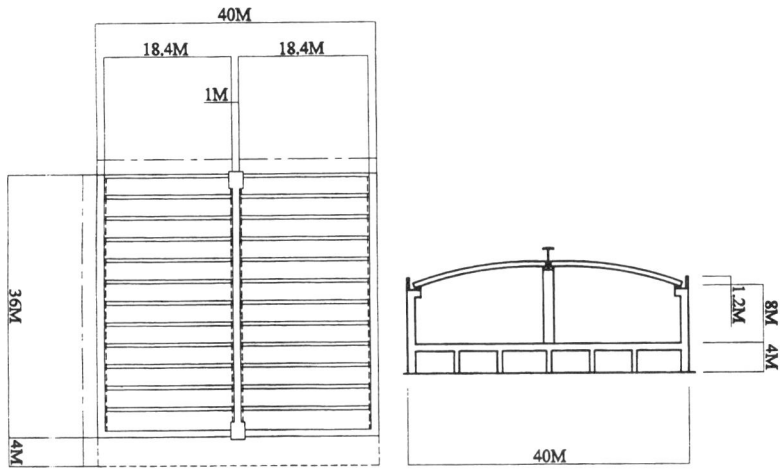

Fig. 9-17. Public gymnasium with a precast beam roof

Fig. 9-18. View of the public gymnasium with a precast beam roof after the earthquake

Fig. 9-19. Bearing connection of another type of the precast beam system

9.3.1.3 Damage to Suspended Ceilings and Suspended Facilities

In the previous subsections structural damages to large-span roofs were already reported; apart from the damage to precast roofs, they were found not to be critical. Our investigation in the Kobe area revealed unexpected damage to many interior spaces of wide-roof buildings. A more severe problem in wide-roof structures is the problem of nonstructural components suspended from roofs, such as suspended ceilings and suspended facilities. Since they are suspended at considerable heights, any damage can easily injure people and spoil the function of internal spaces. Fig. 9-20 shows some examples of damage to suspended ceilings in gymnasiums and event halls. Considerable parts of their ceilings had fallen off during the earthquake.

Ceiling panels are composed of many different materials that are used in many different locations in interior spaces, as shown in Fig. 9-21. Some of them were combined with heavy metal grids while others are plasterboards and still others light glass-wool panels. Some of them are suspended over the arena and some of them cover the areas above the audience seats.

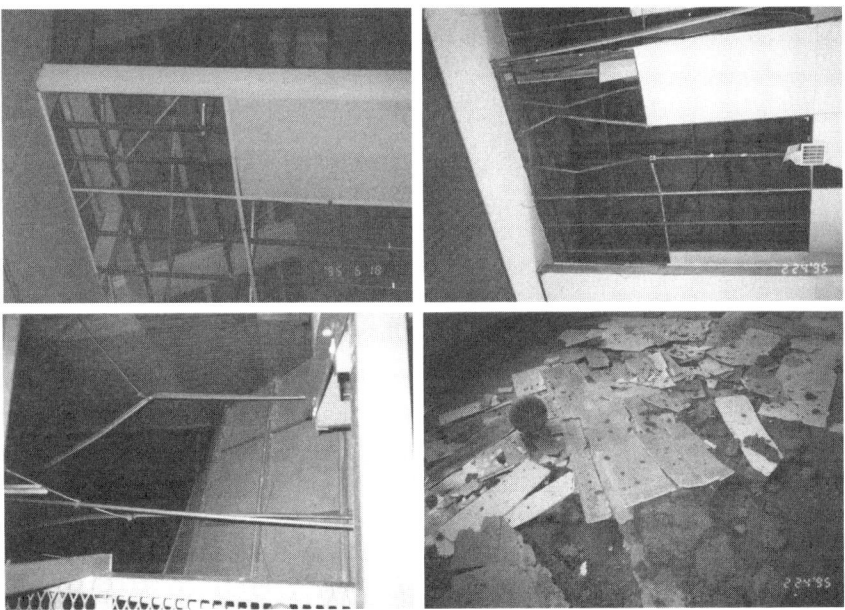

Fig. 9-20. Damage to suspended ceilings

Fig. 9-21. Suspended ceiling panels

There are many things suspended under wide-roof structures: lighting, speakers, air ducts, acoustic panels for theaters or huge video panels for event arenas, for example. Some of them are very heavy and hung in the middle of event halls. Some of them are scattered in the interior space at the ceiling level. Fig. 9-22 shows a huge speaker cluster in a big arena and an acoustic panel in a theater felled by the Kobe earthquake. It has been reported that during the earthquake the winch bearing cables of the speaker cluster failed first, then the cluster freely fell to the length of the cable, bouncing, with finally cut the suspended cable.

Through our investigation in the Kobe area, we have collected data for 117 wide-roof buildings, as shown in Fig. 9-23, including undamaged buildings. In nearly one third of these cases (38 out of 117), failures of suspended nonstructural components were found. In 31 cases, almost seventy percent, the failures were the fall of suspended ceilings. This means that about one third of the wide-roof buildings in Kobe suffered from the fall of nonstructural components, making the suspended ceilings one of the most vulnerable elements among the nonstructural components.

After the Kobe earthquake many large public spaces with wide roofs were converted into refuge stations for those who lost their houses. The importance

Fig. 9-22. Failure of suspended facilities

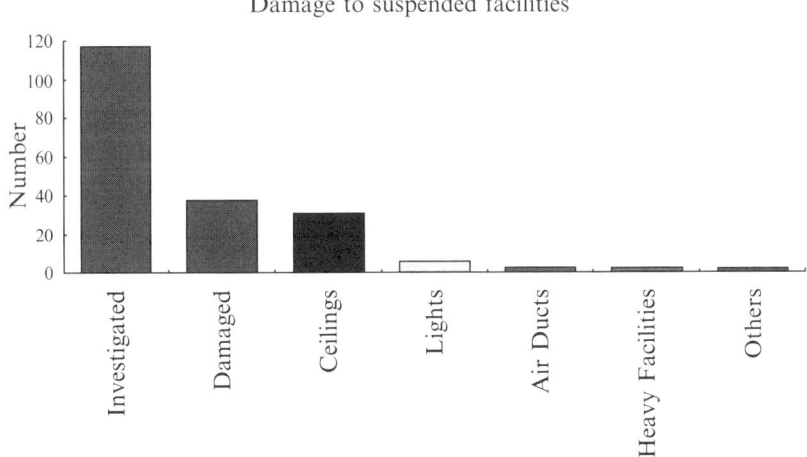

Fig. 9-23. Results of investigation of 117 wide-roof buildings

of the function of wide-roof buildings as refuge spaces was widely dis-
seminated to everybody through mass media. The failures of nonstructural
components can easily blemish this important role of wide-roof buil-dings;
there were actually many public gymnasiums and halls that could not be

opened to the public as refuge stations because of failures of nonstructural components.

9.3.1.4 Summary of the Damage Investigation in the Kobe Earthquake

As a summary of the investigation carried out after the Kobe earthquake, the following points about the damage to wide-roof buildings have been pointed out:

1. Structural damage to wide-roof buildings was generally minor.
2. Failures of vertical braces in supporting structures or horizontal braces in roof frames were commonly observed. However, such damage did not cause any fatal damage to the structures.
3. Damage at the connection between roof frames and supporting structures at bearings was commonly observed. Especially for reinforced-concrete supporting structures, removal of covering concrete at the bearing was often observed. Structurally this is not critical; however, this is dangerous for the people using the interior space.
4. Some member failure is often observed in roof structures with space frames. However, because of high redundancy of the systems, they preserved much leeway before the structure reached the failure mechanisms.
5. A pneumatic dome survived safely, but it suffered from a long-term power failure.
6. Two precast beam roofs designed after the Silberkuhl system collapsed badly. Any such building in any seismic region must be immediately investigated, checked and improved. Precast beam roof system must be designed more carefully.
7. Damages to nonstructural components such as suspended ceilings or suspended facilities have been found to be serious. This poses dangers not only to the people using the interior spaces but also negatively impacts the very important function of large-roofed buildings as refugee stations. More attention must be paid to this type of damage to the interior space of wide-roof buildings, and proper improvements need to be made.

9.3.2 Other Earthquakes After the Kobe Disaster

After the Kobe earthquake, several major earthquakes have occurred in Japan. In each case, public spaces like gymnasium or event halls with wide roofs have provided refuge stations for the people who could not go back to their houses. This function of wide-roof buildings is regarded to be very important. However even minor problem in the nonstructural components, such as slight dislocation of a ceiling panel, may negatively impact this function. People can never feel safe under these circumstances. This is another important characteristic of such problems.

This section discusses the safety of wide-span buildings in terms of damage to nonstructural components. As can be easily predicted from the earlier summary of the investigation of Kobe earthquake, damage to structural components found in other earthquakes after Kobe have been always minor. In contrast, the failure of nonstructural components in wide- roof buildings have caused more problems and attracted more public attention.

Pictures in Fig. 9-24 show some examples of damage to suspended ceiling in wide-roof buildings after various recent earthquakes in Japan. The picture at the top on the left is from the Kobe earthquake. The top picture in the middle is from the Niigataken-Chuetsu earthquake. The top on the right is from the Tokachi-oki earthquake in 2003 and the bottom is from the Miyagiken-oki-no earthquake in 2005.

After the Kobe earthquake, people learned the necessity of structural reinforcement to the buildings designed by old codes because many old buildings were damaged seriously and people were killed by such failures. This was the typical response after Kobe earthquake. Retrofit reinforcement to the old buildings came to be regarded as the most urgent task.

After the investigation of Kobe earthquake I personally warned in many lectures about the danger of nonstructural components in wide-roof buildings. However, the Kobe earthquake occurred early in the morning and very

Fig. 9-24. Examples of damage to suspended ceilings

Fig. 9-25. Miyagi-ken earthquake (Courtesy of Dr. Y. Sakai)

few people were injured in wideroof buildings. Therefore, no serious attention has been applied to this problem. It was not until the Geiyo earthquake in 2001, when people were seriously injured by collapsed suspended ceilings and wall panels, that the problem was publicized and the government realized its urgency.

After the Miyagiken-oki earthquake in 2003, one of the school gymnasiums had provided a refuge station. However, minor dislocations of some ceiling panels were found and the refugee people had to move soon to another gymnasium because of the possible failure of the panels by aftershocks, as shown in Fig. 9-25. This is a typical example of the fact that even minor damage to nonstructural components can negate the function of wide-roof spaces.

During the Tokachiken-oki earthquake in 2003, suspended ceilings in the terminal building and control towers in Kushiro airport failed and fell to the floor in wide scales. The airport became completely unserviceable because of these failures just after the earthquakes, as shown in Fig. 9-26.

9.3.3 Niigataken-Chuetsu and Fukuokaken-Okino Earthquakes

Impacts of the severe Niigataken-Chuetsu Earthquake of 2004 are shown in Fig. 9-27. After the main strike, several major aftershocks subsequently

Fig. 9-26. Damage to the ceiling in Kushiro airport terminal

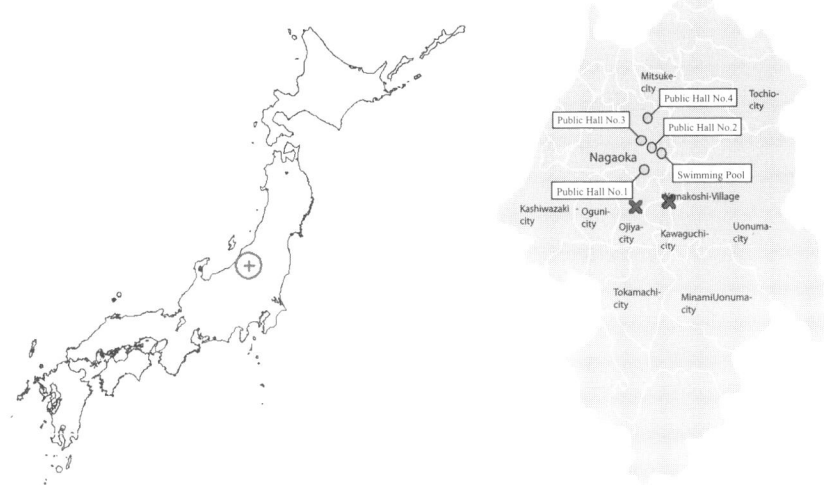

Fig. 9-27. Location of the epicenter of the Niigata-Chuetsu earthquake

followed. Landslides occurred on a considerable scale and many infrastructures were damaged, including roads, railways, and a Shinkansen line.

After this earthquake we also carried out damage investigation for wide-roof buildings. Since the epicenter was close to a local village with fewer wide-roof buildings, our investigation was concentrated in the Nagaoka area, the nearest urban city to the epicenter. The structural damage found in Nagaoka was similar to the damage that we found in Kobe in 1997, but more minor. However, most wide-roof buildings suffered from damage to the suspended ceiling systems, as shown in Fig. 9-28 and Fig. 9-29. The panels in Fig. 9-28 are light and soft glass-wool panels. However. it can also be observed that some of the furring metal members had fallen with the panels. This fact made the damage more dangerous. In one of the buildings shown in Fig. 29, the ceiling panels, consisting of wood-wool cement boards, fell not only to the floor but also onto spectator seats. In this gymnasium a young girl on the floor suffered serious head injuries by these ceiling panels during the earthquake.

In Fig. 9-30, failures of the ceiling panels in a swimming pool are shown. In this case the ceiling panels and other components were set on the flanges of T-shaped linear furring members. During the earthquake the panels and some lighting systems became dislocated from the flanges and fell off.

After the Niigataken-Chuetsu earthquake came the Fukuokaken-okino earthquake in 2005. After this earthquake we also investigated some wide-roof buildings and found some damage to nonstructural components, as shown in Fig. 9-31 and Fig. 9-32.

9.3.4 The Failure in a Swimming Pool

After the Fukuokaken-okino earthquake came the Miyagiken-okino earth-quake in 2005. The earthquake itself was not that severe. But the shocking

Fig. 9-28. Glass-wool ceiling panels in a public gymnasium in Nagaoka city

Fig. 9-29. Damage to ceiling systems in public halls in Nagaoka city

Fig. 9-30. Damage to ceiling systems in a swimming pool

Fig. 9-31. Damage to nonstructural components during Fukuoka earthquake

Fig. 9-32. Damage to facilities settled in the heights during Fukuoka earthquake

news came when 30 people were injured by the falling of ceiling panels in a swimming pool in the Sendai region, as shown in Fig. 9-33. Two days afterwards we visited the swimming pool for to investigate the failure. After collecting the information at the site we had the impression that this particular failure seemed different from other cases seen before. For one thing, this happened just a month after the opening of the swimming pool. The earthquake was not the strongest, though most of the ceiling panels collapsed. Also, the remains of the failure exhibited strong symmetry.

Just by chance I had an opportunity to obtain the image of the security video that showed the very instant of the fall of the ceiling panels. The video shows how quickly the ceiling fell onto people during the earthquake. Almost ninety percent of the ceiling panels fell off, and most of the floor and pools were covered by them. Some people were seriously injured as they punched off the ceiling panels by their heads. News reports quoted people as saying that the ceiling started to fall from one end of the room. However, as seen in the video, it also started to fall from the other end. This proves that the failure occurred almost simultaneously and symmetrically from both ends.

What was the main cause of this particular failure? This is still not clear. Because of the symmetric characteristics of the failure I personally think that the geometry of this particular room and the ceiling system are strongly connected to the main causes. We are still investigating this problem with my students.

Fig. 9-33. Serious failure of ceiling systems in a swimming pool in Sendai

9.3.5 The Cause of the Failures

The ceiling system used in the swimming pool in the previous section was one of the most common and conventional systems. This system uses "clips," or thin plate connections between the furring metal components as shown in Fig. 9-34. It has been said that this clips are one of the weakest points of the system. At the site numerous deformed clips were found. It is almost clear that deformation of the clips finally let the panels move off from the furring components.

The behavior of the clips during earthquakes is becoming clearer through some recently shaking tests. The connection slips during the shaking movement. and the clips gradually bend and finally release their grips. This

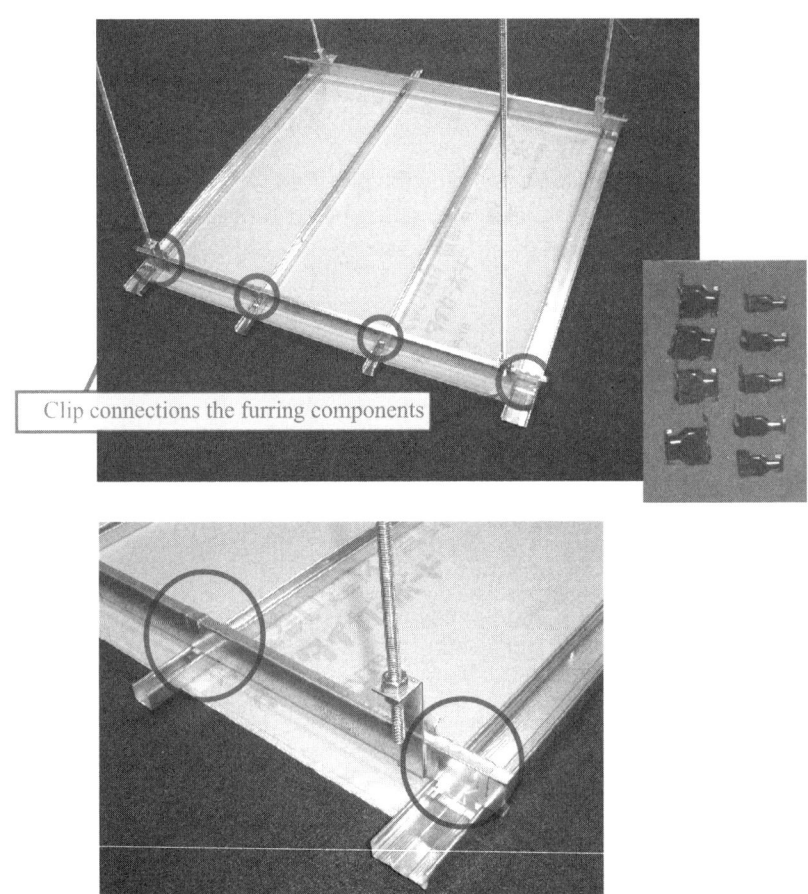

Clip connections the furring components

Fig. 9-34. Connection between ceiling panel furrings and supporting structures

causes local failure of the ceiling system. However it is still difficult to observe a failure on a larger scale—which we have in the real damage—even in the full-scale test by the shaking devices. This means there still exist unknown factors around this phenomenon.

I have seen many damages through my post-earthquake investigations. Many other types of damage can occur in ceiling systems, as is shown in Fig. 9-35. For me, this deformation of clips is therefore just one typical failure of the ceiling systems. One should be aware that even after these connections have been reinforced and improved, still another type of failure could result in the ceiling systems. One can never get rid of damage to non-structural components without reconsidering the total systems.

Fig. 9-36 shows another example of damage of ceiling panels in another swimming pools. In this case, in 2006, most of the ceiling panels fell off not during the earthquake but in the morning of an ordinary day. The failure occurred at the connection between boards and nails that fixed the panels to the furring bars. Perhaps because of the moisture and long-time deterioration of the ceiling panel, the nails' heads came out from the boards.

Fig. 9-35. Various types of failure of ceiling systems

Fig. 9-36. Sudden collapse of ceilings with no earthquake in a swimming pool in Saitama

Table 9-3. Non-earthquake failures of nonstructural components reported in 2004

Date	Place	Facts
2 April evening	Shopping Mall Sao Paulo, Brazil	Plaster ceiling fell. 9 injured
15 May 22:05	Theatre Royal Haymarket London, UK	Plaster ceiling fell. 15 injured
11 October 17:15	Ice-skating rink, Shiga Japan	2 ceiling panels fell off. No injuries. A major skate meeting was stopped
16 December 13:20	Oakland Int. Airport, New Zealand	Ceiling tiles fell off. 15 injured

Surveying the world, one can find many failure reports of the ceilings in wide-span spaces, as shown in Table 9-3. This means the danger of ceiling failure in wide-roof spaces exists not only in seismic areas but in nonseismic areas as well.

For example, in terminal 1 of the Charles de Gaulle-Roissy Airport in Paris, one can observe that nets cover the entire ceiling surface of the terminal in order to prevent the possible removal and fall of the plaster ceilings, as shown in Fig. 9-37.

Fig. 9-37. Ceilings in the Charles de Gaulle-Roissy airport

9.3.6 The Three Conditions

As pointed out in the last part of the previous section, nonstructural components in wide-roof buildings can fail easily, and not only by earthquakes. Failure of nonstructural components becomes extremely dangerous when they have been put at the locations that satisfy the following three conditions.

1. The space is used for large public gatherings.
2. The nonstructural components are placed very high. and
3. The space is very wide.

If these three conditions are satisfied, people using the space would possibly be endangered. not only during earthquakes but also in ordinary time. Furthermore, as has already been mentioned earlier, wide-roof buildings provides space for refugee stations after disasters. One should therefore be very careful about the design of nonstructural components in wide-roof buildings.

9.4 Some Remedies for Preventing Failure of Nonstructural Components

What can then be done to prevent this type of failures? After the failure at Kushiro Airport, another type of ceiling system was employed: one that used membrane material, which seems to reduce the danger very much more so than the previous material, as shown in Fig. 9-38, Fig. 9-39, and Fig. 9-40. Replacing the ceiling by using a lighter and safer system is one way to prevent the failures. However since some metal components are still used, it is yet very difficult to say that it would never hurt the people under any conditions. So far, there exist no criteria for judging and comparing

Fig. 9-38. Kushiro airport terminal

Fig. 9-39. Membrane textile ceilings in the reopened Kushiro airport

the level of safety of nonstructural components. This is another point that should be investigated in the future.

Another option would be not to use ceiling systems at all. For example, the swimming pool at Sendai was reopened without ceiling systems one year after the disaster, as shown in Fig. 9-41. In this case some of the functions of ceilings were lost, such as thermal insulations, acoustics and esthetic effects. but this is still better than to have another failure.

Fig. 9-40. Details of the membrane ceilings

Fig. 9-41. Reopened swimming pool without ceilings

Finally, as seen in the example of Charles de Gaulle-Roissy Airport, covering the existing ceiling surface with nets is also another effective option for preventing failure, although it may not be a permanent solution.

9.5 Conclusions

This article began with a discussion of the seismic performance of large enclosures and then continued with a discussion about the damage to non-structural components. Illustrated in the introduction were many types of failure of wide-roof buildings in disasters. In the first part the main focus was put on earthquake damage; in the second part the stress was put on the failure of the nonstructural components.

The damage investigation after the Kobe earthquake revealed the hidden dangers of the damage of nonstructural components. The function of wide-roof buildings as refugee stations was also recognized. Later investigations gradually deepened the understanding of the damage of the nonstructural components. Vulnerability of non-structural components must be regarded as a potential threat if the space satisfies the three conditions mentioned in Section 9.4. Nonstructural components should be designed very carefully if the space fulfills these three conditions.

How to prevent the failure of nonstructural components in wide-roof buildings is a very urgent problem. Three types of remedies are herein indicated: replacement with a safer system using light materials, removal of the ceiling system, and putting nets under the ceilings. In order to establish guidelines for the design of nonstructural components, criteria are needed to evaluate the safety of the systems.

Safety of the nonstructural components is not a conventional structural problem and cannot be solved by conventional approaches. This aspect of building safety is very important and should be carefully considered for wide-roof buildings. Safety cannot be realized only by structural engineers. First of all, it is necessary for architects to have an understanding of these issues. Then it is essential for there to be cooperation between architects and engineers.

10. Antiseismic Design: From Houses to Large Enclosures Wooden Houses

Mikio Koshihara

10.1 Introduction

Japan has a longstanding heritage of wooden houses. At the same time, Japan is a country beset by earthquakes. Wooden houses in Japan have suffered great damages caused by strong earthquakes. In the 1995 earthquake in Kobe, a great number of wooden houses was destroyed and more than 5,000 human lives were lost. This paper presents an overall survey of wooden houses mainly in Japan, focusing on their seismic performance.

10.2 Categories of Wooden Houses in Japan

As shown in Table 10-1, there are many types of wooden buildings in Japan. They are different in terms of seismic elements and how each seismic design method is applied.

10.2.1 Traditional Wooden Buildings

The first category, traditional wooden buildings, includes temples, shrines and old folk houses (Fig. 10-1). However, at present very few are newly constructed of wood.

10.2.2 Detached Wooden Houses

The second category includes single-family detached wooden houses (Fig. 10-2). Japan has a stock of about 30 million units, and 600 thousand units are newly constructed every year. This category can be divided into

Y. Fujino, T. Noguchi (eds.) *Stock Management for Sustainable Urban Regeneration*,

Table 10-1. Categories of wooden buildings and design method

Seismic design Type of buildings and construction		Seismic elements	Seismic design method
Traditional wooden buildings	Shrine, temple, pagoda Folk house	Frame, wall and rocking effect	Not applicable or special analysis
Detached wooden houses	Conventional two-by-four Prefabricated	Shear wall (bracing or board)	Effective wall length method
	Moment-frame system		Special analysis
Heavy timber structure	Ordinary frame	Moment resisting frame, truss, etc.	Working stress (allowable stress) design method
	Long span or space frame		
High-rise wooden building		Moment resisting frame, shear walls, etc.	Performance- based design method

four types according to the classification of construction method: conventional construction, two-by-four system, prefabricated houses and moment resisting frame system.

The conventional construction method was derived from traditional construction methods (Fig. 10-2(1)). The structural system of both methods has posts and beams. This type of houses account for the largest share (60 percent) among wooden detached houses.

The two-by-four system (Fig. 10-2(2)) was introduced from North America, where its original system is the most common construction method

(1) Traditional Wooden Building   (2) Traditional Wooden Houses

Fig. 10-1. Traditional wooden buildings, (1) Traditional wooden building, (2) Traditional wooden houses

and is called light-frame construction or platform construction. The name "two-by-four" comes from the sectional size of dimension lumber (approximately two inches by four inches). This structural system consists of walls with nailed boards.

There are various systems of prefabricated wooden houses, but the most popular system is a panelized one (Fig. 10-2(3)). Walls, roofs, and floors

(1) Conventional construction   (2)Two by four Houses system

(3) Prefabricated Houses   (4) Moment frame system

Fig. 10-2. Detached wooden houses, (1) Conventional construction, (2)Two by four houses system, (3) Prefabricated houses, (4) Moment frame system

(1) Heavy Timber Structure   (2) 5 storied Wooden Building

Fig. 10-3. Heavy timber structures, (1) Heavy timber structure, (2) 5 storied wooden building

1995 Hyogoken-nambu Earthquake   2004 Niigataken-chuetsu Earthquake

Fig. 10-4. Damage to wooden houses in Japan, 1995 Hyogoken-nambu earthquake, 2004 Niigataken-chuetsu earthquake

are made of wood panels with glued boards, and its structural system is based on these panel elements.

Recently some houses have been constructed with a moment-resisting frame (Fig. 10-2(4)).

10.2.3 Heavy Timber Structure

The third category is heavy timber structure, used for example, to build schools, gymnasiums and museums (Fig. 10-3(1)). After around 1960 this type was replaced by steel structure. However, quite a few heavy timber buildings have been built after around 1980.

There are great varieties in the structural system of heavy timber structure. Moment- resisting frames and trusses are widely used for ordinary buildings such as schools, and space frames are often employed for big buildings such as gymnasiums.

Since 2000 we have been able to build fireproof wooden structures. For example, a new five-story wooden building has been built in Kanazawa (Fig. 10-3(2)). This building was built by a wooden hybrid system whose elements, columns, and beams are wood-inserted steel bars or steel plates.

10.3 History of Damages Due to Earthquakes

Many earthquakes have occurred in and around Japan, and many typhoons have also besieged her in the past. Some of them resulted in a great amount of damage to wooden buildings, with many human lives lost. Table 10-2 shows such disastrous earthquakes and typhoons.

Table 10-2. Major disastrous earthquakes in Japan (After Meiji restoration)

Year	Earthquake and typhoon	Magnitude	Damage	Casualty
1981	Nohbi earthquake	8.0	220,000	7,273
1923	Kanto earthquake	7.9	254,000	142,000
1934	Muroto typhoon	–	93,000	2,702
1948	Fukui earthquake	7.1	48,000	3,769
1964	Niigata earthquake	7.5	9,000	26
1968	Tokachi-oki earthquake	7.9	4,000	52
1978	Miyagiken-oki earthquake	7.4	7,000	28
1995	Hyogoken-nambu earthquake	7.2	240,000	6,433
2004	Niigataken-chuetsu earthquake	6.8	17,000	65

The 1981 Nohbi Earthquake, which occurred in the middle of the main island of Japan, was a very big one with estimated magnitude of 8 on the Richter scale. Most wooden buildings at the time were built by carpenters using traditional construction methods. After the earthquake, scientific research on the seismic design of buildings were initiated.

Next in intensity was the 1923 Kanto Earthquake that hit Tokyo and the metropolitan area of Japan. The disastrous earthquake caused great damage to buildings as well as loss of human lives. The first regulations on seismic design were introduced into Japanese building law after this earthquake: a seismic coefficient of 0.1 for seismic design was adopted. However, only the following stipulation was added at the time: that some bracings should be installed in wooden buildings. Accordingly, most wooden houses did not install such bracings.

The 1948 Fukui Earthquake occurred in a rural district, but many people lost their lives mainly due to the collapse of detached wooden houses. The investigation of the damage showed that the degree of the damage to wooden houses had a strong correlation with the shear wall ratio of the house. The wall ratio is defined as the amount of walls in length in the unit area of the floor. The result was taken into consideration when the new Building Standard Law was established in 1950.

The Niigata, Tokachi-oki, and Miyagiken-oki earthquakes were big ones that resulted in great damage to buildings, especially reinforced concrete structures. Wooden houses also suffered damage, but the damage was not so serious. It looked as if Japanese wooden houses had finally become strong enough to survive severe earthquakes.

However, such an optimistic view was betrayed by the 1995 Hyogoken-nambu Earthquake, which is also known as the Kobe Earthquake. About

100,000 buildings, including wooden houses, were completely destroyed (Fig. 10-4) and more than 5,000 people were killed. It is estimated that about 80 percent of them lost their lives because of the collapse of wooden houses. Various investigations and researches have been made after the earthquake. Some results were taken into account in the revision process of the Building Standard Law in 1998. After the Kobe Earthquake, many disastrous earthquakes have occurred.

10.4 Detached Wooden Houses

There are many wooden detached single-family houses in Japan; there is a stock of about 24.5 million units, with half a million being newly constructed every year. This category can be divided into four types according to the classification of the construction method (Fig. 10-2):

(1) Conventional Method of Construction

The conventional construction method was derived from the traditional construction method. The structural system of both methods has posts and beams. Houses of this type represent the largest share (60 percent) among wooden detached houses.

(2) Two-by-Four System

The two-by-four system was introduced from North America, where the original system is the most common construction method and is called light-frame construction or platform construction. The name "two-by-four" came from the sectional size of dimension lumber, in inches. This structural system consists of walls with nailed boards.

(3) Wood Panel Construction (Prefabricated)

This system, light-frame construction, prevails throughout the world. There are various systems of prefabricated wooden houses, but the most popular system is a panelized one in which walls, roofs and floors are made of wood panels with plywood, and with a structural system based on these panel elements.

(4) Moment-Resisting Frame and Truss

This type of house is designed by structural engineers. There are great varieties in structural systems, moment-resisting frames, trusses, wooden shells and so on. In these houses, engineered wood, glued laminated wood, and LVL (Laminated Veneer Lumber), are used as column and beam.

10.5 Seismic Design Methods for Wooden Buildings

10.5.1 History of Design Methods

The provision of seismic design for structures was introduced in Japanese building law for the first time in 1924, right after the great Kanto Earthquake. However, detailed design method for wooden buildings was not introduced until 1950, when the Building Standard Law was established. Since then, the allowable-stress design method is applied to ordinary wooden buildings and the effective wall-length method is applied to detached small wooden houses.

The effective wall-length method is still used today for its simplicity. The numerical values of the coefficient used in this method have changed several times, and the numerical values at present were stated in the revision of the Building Standard Law in 2000. The load-carrying capacity method was introduced in the 1980 revision. However, it had rarely been used for wooden buildings, because the number of large wooden buildings that should be designed by the method was small.

The performance-based design method was introduced into the Building Standard Law in 1998, under the heading of "Calculation of Response and Limit Strength."

10.5.2 Effective Wall-Length Method

The effective wall-length method consists of two components: the required ratio of wall length and the resistance factor of shear wall.

The required ratio of wall length, p, was determined based on the assumption that the weight of each part (roof, wall, and floor) is the same in typical wooden houses and that the design base shear coefficient is 0.2 according to the minimum requirement in the Building Standard Law. The value of resistance factor of a shear wall, q, was determined based on racking tests. The following relationship should be satisfied in the seismic design of every wooden house.

$$p \times A \leqq \Sigma(q \times l)$$

where
  p = required ratio of wall length (cm/m^2, Table 10-3)
  A = area of the floor (m^2)
  q = resistance factor of each shear wall (Table 10-4)
  l = real length of shear wall (cm)

Table 10-3. Required ratio of wall length $p(\text{cm/m}^2)$

Story Roof	Single storied	Two storied		Three storied		
		1st story	2nd story	1st story	2nd story	3rd story
Heavy (Clay tile)	15	33	21	50	39	24
Light (Metal plate)	11	29	15	46	34	18

Table 10-4. Resistance factor of wall q (examples)

Type of wall		Resistance factor
Mud wall		0.5
Bracing (mm)	15×90	1.0
	30×90	1.5
	45×90	2.0
	90×90	3.0
Plywood t $\geq$ 12 mm		2.5
OSB t $\geq$ 12 mm		1.0
Gypsum board t $\geq$ 12 mm		1.0

In the above expression, $p \times A$ corresponds to the earthquake force and $\Sigma(q \times l)$ corresponds to the resistance.

In the revision of the Building Standard Law in 2000 two design techniques were introduced clearly. One is the design technique for the effect of eccentricity and the other is the design technique for the joint of the top and bottom of columns with wall.

In Japan wooden detached houses have been designed not only by structural engineers but also carpenters. This is why the very simple design method of effective wall length is needed.

10.5.3 Allowable-Stress Design Method

The allowable-stress design method has been used for bigger buildings, such as schools and offices. Recently this method has been used for wooden detached houses, too. There are moment-resisting frames as seismic elements, but these are not included as seismic elements in the effective wall-length method. The standard seismic base shear coefficient is 0.2. The stress in the members and joints are checked for long-term and short-term loads.

10.5.4 Load-Carrying Capacity Method

In the revision of the Building Standard Law in 1980, load-carrying capacity method (the method used to examine the performance against very severe earthquakes) was introduced. The method is based on the approximate relationship between the linear response and the nonlinear response. In any approximation, the energy of linear response and that of nonlinear response are regarded equal. The seismic base-shear coefficient for very severe earthquakes is 1. The design seismic base shear coefficient could be decreased according to the ductility of the structure.

10.5.5 Performance-Based Design Method

The performance-based design method ("Calculation of Response and Limit Strength") introduced in the recent revision of the Building Standard Law in 1998 is based on the approximate relationship between the linear response and nonlinear response, in which the nonlinear response is replaced by the equivalent linear response. So far, very few wooden buildings have been designed by this method, because it is not yet popular among structural engineers and the data for it are not enough for practical use.

10.6 Seismic Evaluation and Reinforcement Method for Wooden Houses

10.6.1 Seismic Problem for Existing Wooden Houses

As of 2003, there are 24.5 million detached houses in Japan, with half a million more being built every year. In Tokyo alone there are five million wooden houses. Among these existing houses it is thought that 10 million have poor seismic performance. It has been estimated that 90 percent of the houses built before 1981, when the Building Standard Law was revised, have poor seismic performance. These houses were termed existing ineligible houses.

There are two problems related to the seismic performance of existing wooden houses. One is the change in the Building Standard Law: seismic design methods for new wooden houses and the numerical values of co-efficients have changed several times after various disastrous earthquakes (Table 10-5). So from the viewpoint of the building stock, wooden houses are not always designed in adherence to present seismic design methods. The other problem faced by wooden houses is the deterioration wrought by aging and termites.

In the 1995 Kobe Earthquake, 80 percent of deaths were related to old, small houses. It is thus required to make a diagnosis of existing houses and to reinforce the house as the result of this research. Many local governments promote seismic diagnosis and seismic reinforcement in Japan. However, many owners don't request diagnosis and reinforcement for their houses. Many are skeptical of the government's warning that a great earthquake could strike the house during their lifetimes. If the possibilities of strong earthquake hits were given in stochastic value, the meaning of the value would be vague and unreliable. In addition, it would entail a great effort to settle elderly people into a new house. The practice of seismic reinforce-ment would thus depend on the philosophy of owners.

There are many issues involved in trying to promote the reinforcement of existing wooden houses. Seismic engineers must present a reliable and economical seismic diagnosis plus reinforcement methods for existing wooden houses. Some simplified diagnosis method like the effective wall-length method is required for carpenters and house builder who diagnose houses. The seismic-evaluation method of existing wooden house was revised in 2004, and many other reinforcement methods are being developed.

10.6.2 Seismic-Evaluation Method for Wooden Houses

A seismic-evaluation method was published by the Japan Building Disaster Prevention Association and fully revised in 2004. The new Building Standard Law revised in 2000 was linked to this method, in which the performance

Table 10-5. Change of required ratio of wall length (cm/m²)

| Year | Heavy roof | | Light roof | |
	1st story	2nd story	1st story	2nd story
1950	16	12	12	8
1960	24	15	21	12
1891	33	21	29	15
2000	33	21	29	15

of ground, foundation, and building was separately evaluated because the ground disaster cannot be avoided by reinforcing the houses. This seismic evaluation method consists of two components: required strength and resistance strength. This method is almost the same as the effective wall-length method for new houses but it is different in that the value compared is not the length of walls but the strength of walls.

10.6.3 Reinforcement Method for Wooden Houses

It was estimated that 90 percent of the houses built before the 1981 revisions of the Building Standard Law manifest poor seismic performance. In fact the diagnosis point of most existing wooden detached houses built before 1981 is under 0.7. For these houses with poor seismic performance, many techniques of reinforcement have been developed:

(1) Ground

The house standing on a deep alluvium is in danger of liquefaction (Fig. 10-5(1)), but avoiding liquefaction is very expensive compared to the cost of building a detached wooden house. The house on the cliff and under the cliff is in danger of landslide (Fig. 10-5(2)); in this case it is effective to reinforce the retaining wall.

(2) Foundation

In case of insufficient foundation, such as one without reinforced concrete, or concrete blocks and stones, there is danger of cracks, corn-type failure and bending failure of a foundation in an earthquake (Fig. 10-6(1)). To reinforce the insufficient foundation, a reinforced concrete foundation is placed along the existing foundation with shear connectors (Fig. 10-6(2)).

(1) Liquefaction   (2) Land slide

Fig. 10-5. Damages of ground, (1) Liquefaction, (2) Land slide

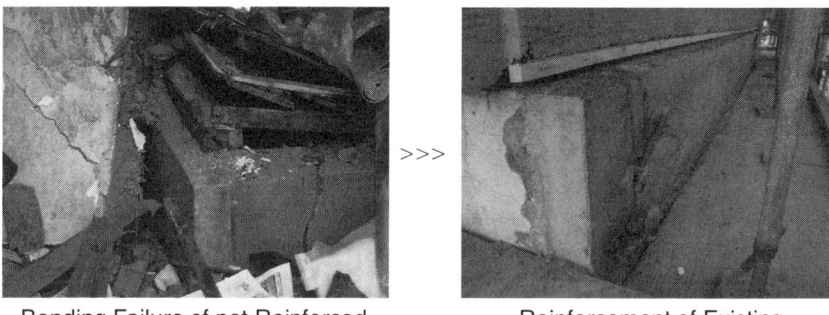

Bending Failure of not Reinforced
Concrete Foundation

Reinforcement of Existing
Foundation

Fig. 10-6. Foundation Bending failure of not reinforced concrete foundation,
Reinforcement of existing foundation

Fig. 10-7. Lack of wall

(3) Lack of Wall

Houses lacking a load-bearing wall suffer huge damage in earthquakes. These houses must install sufficient bearing walls, bracing,
and nailed plywood (Fig. 10-7).

(4) Insufficient Joints

Houses with insufficient joints around the bearing walls are greatly
damaged in earthquakes, even if there is sufficient length of an effective
wall. Insufficient joints around the bearing wall will cause the performance of the wall to decrease because it cannot resist the tensile force of
the column during an earthquake. In this case, fasten the column to the
groundsill and to beams with metal connectors (Fig. 10-8).

(5) Eccentricity

In a small detached wooden house, many bearing walls are arranged
on the northern part of the house and few bearing walls are arranged
on the southern part of house for greater exposure. This causes a large

Fig. 10-8. Insufficient joint around the bearing wall

Damage coused by Eccentricity   Steel Bracing Bearing Wall

Fig. 10-9. Eccentricity problem

eccentricity to the house. In this case, more seismic elements should be arranged on the southern part of the house. New seismic elements such as steel bracing (Fig. 10-9) and rigid frames can add both higher seismic performance and increase the transparency of the wall.

(6) Deterioration and Aging

Wooden building cannot avoid the aging and deterioration brought by using the material, wood. The deterioration on the bottom of column is as same as insufficient joint of column to sill from the viewpoint of structural performance. In this case, the deteriorated elements and the causal factors should be removed (Fig. 10-10).

(7) Others

Many reinforcement methods for existing wooden houses are developing.

Fig. 10-10. Deterioration

Fig. 10-11. Shaking table test of existing wooden houses (Right: not reinforced/ left: reinforced)

The seismic performance of these reinforcement methods was verified by a full-scale shaking table test. In 2004 two houses built 30 years ago were moved and installed as specimens on the shaking table at E-Defense in Hyogo prefecture. Though the two houses were built at the same time and by the same methods, the one on the left (in Fig. 10-11) was reinforced by installing bracing and nailed plywood, with the columns fastened to a horizontal member. The one on the right was left as is. Two test specimens were shaken by the wave observed at JR-Takatori station in the Kobe Earthquake. After the test, the non-reinforced house collapsed but the reinforced one did not.

10.7 Conclusions

New wooden houses can show good seismic performance, and there are useful reinforcement methods for existing wooden houses that have insufficient support. But these have not been widely implemented because of a low sense of popular fear about an earthquake crisis. It is important to cultivate owner attitudes about the dangers of earthquake to insure the safety of housing stocks.

11. Integrated Earthquake Simulation of Earthquake Hazard and Disaster

Muneo Hori

11.1 Mechanism of Earthquakes

11.1.1 Earthquakes in Geological Length Scale

An earthquake is failure of the Earth's crust. As an example, we take a look at the Japanese Islands, which are located near boundaries of the four plates; namely, the Pacific Ocean Plate, the Philippine Sea Plate, the Eurasia Plate and the North America Plate. These plates are moving at the speed of a few centimetres per year in different directions, and mismatch in their displacement generates stress in the plate boundary or inside of the plates. Earthquakes are caused as the failure of some part of the plate boundaries or of the plate itself when the accumulated stress reaches a critical value.

On the vertical cross section, we observe that the Pacific Ocean Plate subducts itself under the Eurasia Plate. The Pacific Plate is bent like a beam, and stress state is tensile or compressive in the upper or lower part of the plate slab, respectively. Earthquakes take place in these places. More important is shearing stress that acts on the boundary between these two plates. This shearing stress is the cause of earthquakes of largest size when a large area of the plate boundary is broken. An earthquake of this kind is called an interplate earthquake. An earthquake that takes place in the plate is called an intraplate earthquake.

11.1.2 Strong Ground Motion and Seismic Response of Buildings

Earthquake waves that are emitted from a source fault propagate in the crust, which consists of several geological layers. The amplitude increases and the

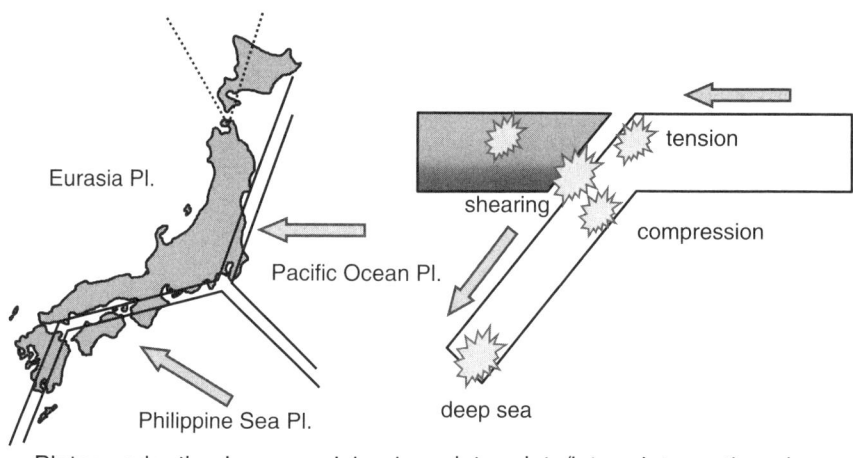

Eurasia Pl.

Pacific Ocean Pl.

Philippine Sea Pl.

tension

shearing

compression

deep sea

Plates under the Japanese Islands   intra-plate/inter-plate earthquake

Fig. 11-1. Earthquake viewed in geological length scale

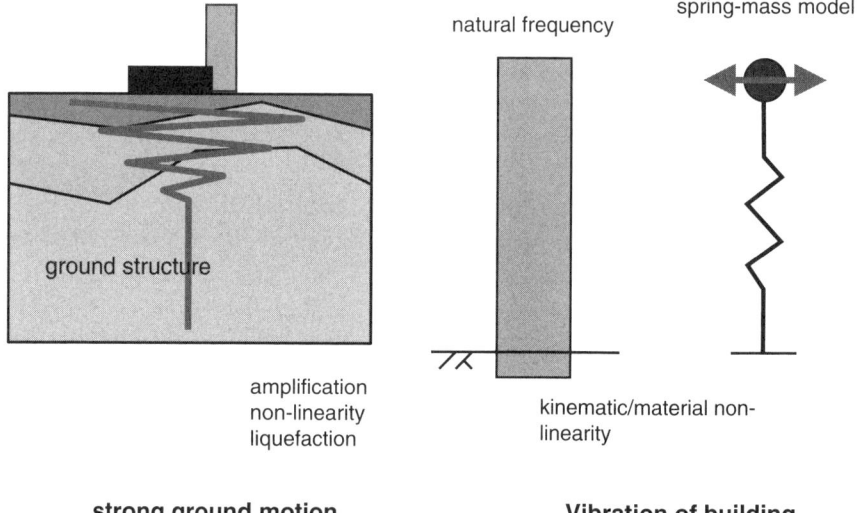

natural frequency

spring-mass model

ground structure

amplification
non-linearity
liquefaction

kinematic/material non-
linearity

strong ground motion   **Vibration of building**

Fig. 11-2. Earthquake viewed in engineering length scale

direction becomes upward as the waves pass through softer layers. When the waves reach the ground surface, the waves are amplified significantly due to soft ground layers near the surface, and their direction becomes almost vertical. Such amplified earthquake waves of the ground surface are called ground motion or strong ground motion in seismology. The strong ground motion often induces liquefaction of soil layers.

A building or a structure has its own natural frequencies; such a human-made structure is stiff and behaves linearly subjected to ground motion. When an earthquake strikes a city, each building or structure is shaken by strong ground motion generated at its site. It usually vibrates at its natural frequency. When the seismic resistance of the building or structure is smaller than the acceleration caused by the ground motion, some parts are damaged and the response becomes non-linear. Damage may increase as the building or structure is being vibrated, which may lead to complete collapse. This is the basic mechanism of structure damage due to earthquake.

11.1.3 Source Process and Strong Ground Motion

In principle, we can predict strong ground motion at any site by calculating the source process in which earthquake waves are emitted from a fault, the propagation process in the crust, and the amplification process near surface ground layers. Besides the fact that there is limitation in modelling underground structures due to the lack in physical survey of the crust and surface layers, it is difficult to predict the source process; for instance, we are not able to measure stress state and strength distribution for faults that are located deep in the crust. We thus give a certain scenario to a source process assuming the location and orientation for a fault.

The distribution of strong ground motion is computed for a given earthquake scenario when a suitable model is constructed for underground structures. The uncertainty of the fault process leads to several or numerous scenarios, and hence there are some ranges in predicting the strong ground motion for a possible earthquake. The range could be large if the uncertainty of the fault process is high.

While we have to make efforts in finding a more certain scenario of a possible earthquake, for the preparation of earthquake disasters, we take for granted that there are many cases for damage distribution of buildings and structures. Also, a case where only a few buildings or structures are slightly damaged might be included. This is one difficulty in promoting earthquake preparedness. We will study how to deal with this difficulty in the next section.

11.2 Necessity of Predicting Possible Earthquake Hazards and Disasters

11.2.1 Earthquake Hazards

Earthquake hazards are strong ground motions caused by an earthquake. The amplitude of the motion changes depending on the earthquake as well as the path along which the earthquake waves propagate and the ground surface layers in which the waves are amplified. It is not possible to clarify the source process of the fault in detail and to survey the geological layer structures. The reliability and accuracy of predicting earthquake waves in the geological length scale are still limited.

Local topographical effects of the ground layers on the amplification cannot be underestimated; the seismic index of JMA often changes by 1 in a distance of few hundred meters or less. By using data of a boring hole and analyzing the wave propagation near the ground surface, the amplification characteristics of ground at a target site can be determined. However, such surveys as use boring holes are made only for large or important buildings and structures.

11.2.2 Earthquake Disasters

Earthquake disaster involves damage to or collapse of buildings and structures due to strong ground motion. Such damage takes place if the seismic resistance of the buildings and structures is smaller than the force that is caused by the strong ground motion.

Prediction of earthquake disaster is not an easy task. It is difficult to predict the strong ground motion that is input to a building or a structure of interest, due to the spatial variability of the strong ground motion. Estimation of the seismic resistance is not simple for existing buildings and structures. The materials might have deteriorated in long-time use, and the resistance tends to gradually decrease. Such loss in the seismic resistance is taken into consideration when buildings and structures are designed. Still, unpredictable loss of seismic resistance may happen, and there are some uncertainties in the seismic resistance for existing buildings and structures.

11.2.3 Risk Analysis

According to the framework of risk analysis, we first estimate risk with uncertainty. What is risk for an earthquake? It is earthquake disaster that consists of direct and indirect loss of property value, human life or limb, or economic activities. The direct loss of estimates is the term used to describe damage to buildings or structures that is caused by strong ground motion when their seismic resistance is not sufficient. The indirect loss is mainly related to the loss of economic activities due to the malfunctioning of social infrastructures. This is the consequence of the direct loss of earthquake disaster.

The risk analysis of earthquakes starts from the estimation of earthquake disaster. The estimation, however, is not easy since there are many unknowns for a possible earthquake: the source process is not predicted or geological structures may not be fully surveyed (even surface ground layers have some ambiguity due to the limitation of boring holes). Still, we may make a decision that we should take some actions such as retrofitting buildings and structures or using financial measures such as earthquake insurance. Decision-making and taking actions are key components of the risk management that is based on the risk analysis of earthquakes.

11.2.4 Social Need for Preparing for Large Earthquakes

The risk analysis has been developed in the field of economics, more related to microeconomic than to macroeconomics. However, it is still quite difficult to apply ordinary risk analysis methods to earthquake disaster, because a large earthquake often results in regional or nationwide catastrophic disaster that spatially and temporally propagates due to the interdependency among social infrastructures.

There is strong social need for preparing for large earthquakes. Constructing new seismic-proofed structures and retrofitting existing structures to a sufficient degree are countermeasures to prevent earthquake disasters. These countermeasures are not cheep. In order to make rational preparation, according to the risk management scheme, we should start from recognizing the possible hazard and disaster that could happen in a target city or region when a large earthquake occurs. Therefore, what is needed is technology that makes it possible to predict earthquake hazards and disasters.

11.3 Development of Integrated Earthquake Simulation (IES)

11.3.1 Limitation of Statistical Analysis

The standard methodology of predicting earthquake hazard and disaster makes use of statistical analysis of past events. An index of strong ground motion is computed by using the so-called empirical equation, which is obtained by analyzing data of measured strong ground motion and source earthquakes. An index of building and structure damage is computed in a similar manner when data of structures damaged by past earthquakes are statistically analyzed.

The accuracy of such statistical analysis is not high. Each earthquake has distinct properties in the source process or the wave-propagation process. Damaged buildings and structures have particular seismic resistance that are not sufficient to withstand the degree of input strong ground motion.

The alternative to this statistical analysis is methodology that is based on numerical simulation. Since digitized information is being accumulated and becoming available, we are able to construct an analysis model for a town or a city. When a model is made, various numerical analysis methods can be applied to predict possible hazard and disaster for a given scenario of earthquake. Although this methodology needs large-scale numerical computation, the accuracy of the prediction will be higher than the standard statistical analysis. Methodology based on numerical simulation is applicable to a rapidly changing city or town if a suitable model is provided for it. The reliability of the prediction depends on the quality of the analysis model, since advanced numerical analysis methods that are developed for a design purpose are available.

11.3.2 Overview of IES

Integrated earthquake simulation (IES) is a system that consists of earthquake simulation, structure response simulation, and action simulation (Fig. 11-3). The earthquake simulation computes generation and propagation of earthquake waves for a given earthquake. Response and damage to a structure subjected to strong ground motion is computed in the structure response simulation. At last, the action simulation computes human or social actions against earthquake disasters. The basic characteristics of these numerical simulations are summarized as follows:

1. The earthquake simulation: Earthquake waves are synthesized according to a scenario of possible fault processes. The propagation of the waves through the crust is computed, and the amplification of the wave near the ground surface is calculated by accounting for the three-dimensional topographical effects and the non-linear properties of surface soil layers.
2. The structure response simulation: Seismic responses of all structures located in a target area are computed. Structures range from residential buildings, infrastructures (concrete, steel, or geotech-

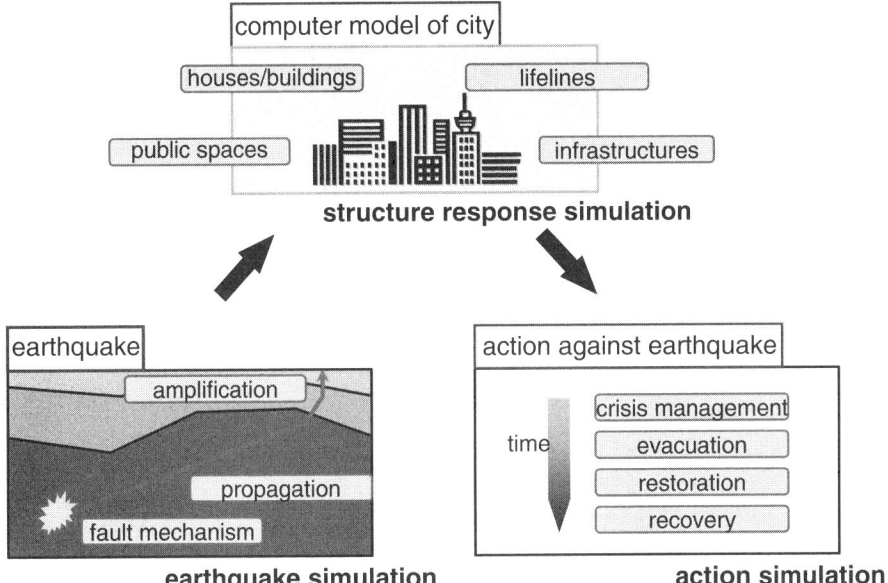

Fig. 11-3. Overview of IES

nical structures), and energy, transportation, communication or water-supply lifelines. Since there are various kinds of structures, a suitable numerical analysis method should be chosen depending on the type of structure.

3. The action simulation: Human behavior is simulated, including evacuation from damaged structures, crisis management, and the analysis of recovery plans.

While each simulation has its own purpose, the three simulations are related to each other, i.e., the earthquake simulation provides strong ground motion distribution to the structure response simulation, with the strong ground motion at its site used as an input wave for each building. Structural damage computed by the structure response simulation provides an initial condition for the action simulation.

The key elements of IES are the geographical information system (GIS) and the three groups of numerical simulations (Fig. 11-4). GIS provides data for constructing computer models, i.e., underground structural data for the earthquake simulation and structural data for the structure response simulation. Results of the numerical simulation are stored in GIS; the earthquake simulation and the structure response simulation provide predictions of earthquake hazards and disasters, respectively, for a given earthquake scenario.

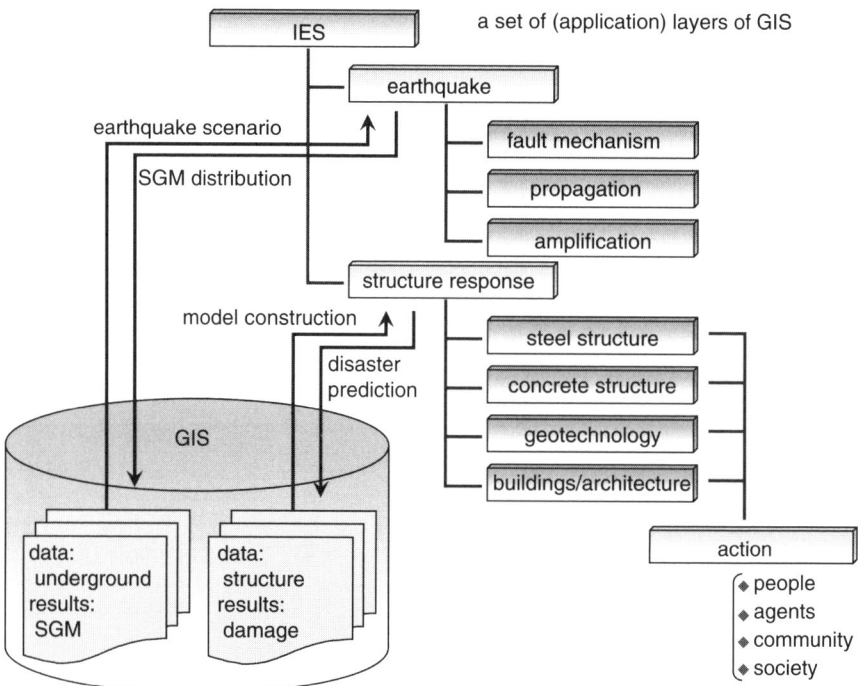

Fig. 11-4. System of IES

11.3.3 Earthquake Hazard Prediction of IES

The major feature of IES is that it offers the highest spatial resolution in computing the distribution of strong ground motion in a target city (Fig. 11-5). This computation is made by taking advantage of bounding medium theory and multiscale analysis. The bounding medium theory provides pessimistic and optimistic underground structure models when only limited information is given to the underground structure of the target city. The multiscale analysis is for efficient numerical computation; a numerical solution computed in a large length scale is refined by analyzing a focused area in a small length scale.

As an example of earthquake hazard prediction using IES, we present the results of the Yokohama City simulation (Fig. 11-6). An actual earthquake of August 11, 1999 is synthesized, and the synthesized waves are compared with the data observed at 12 seismograph sites located in the city. The underground structure models used for the computation are presented

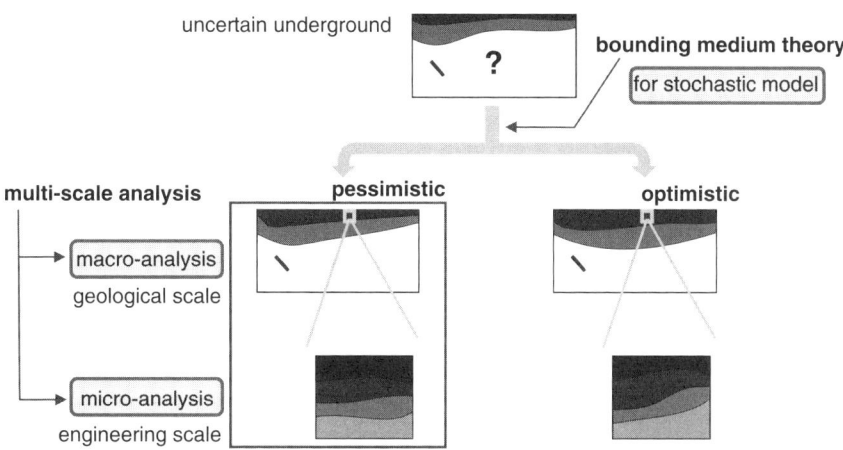

Fig. 11-5. Macro-micro analysis for strong ground motion prediction

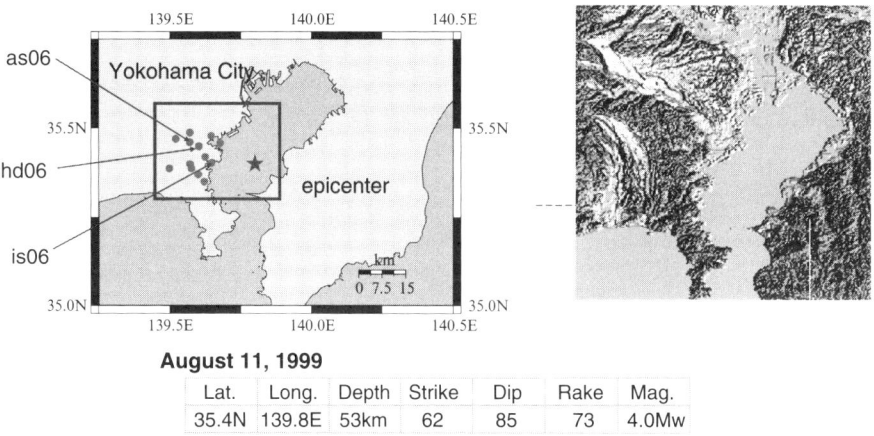

August 11, 1999

Lat.	Long.	Depth	Strike	Dip	Rake	Mag.
35.4N	139.8E	53km	62	85	73	4.0Mw

Fig. 11-6. Example of macro-micro analysis: Yokohama city simulation

in Fig. 11-7 and Fig. 11-8; Fig. 11-7 is for the underground structure in the geological length scale while Fig. 11-8 is for the surface ground-layer model in the engineering length scale.

The waveforms of the synthesized waves are compared in Fig. 11-9. While a low-pass filter of 2.0 Hz is applied and higher frequency components are excluded in the synthesized and observed waves, the agreement is satisfactory. The Fourier spectrum of displacement is compared in Fig. 11-10. It is seen that

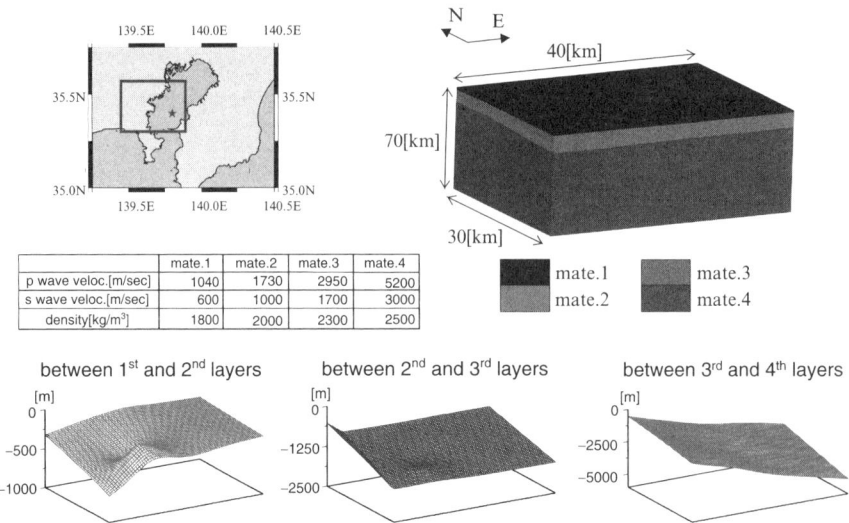

Fig. 11-7. Model for macro-analysis

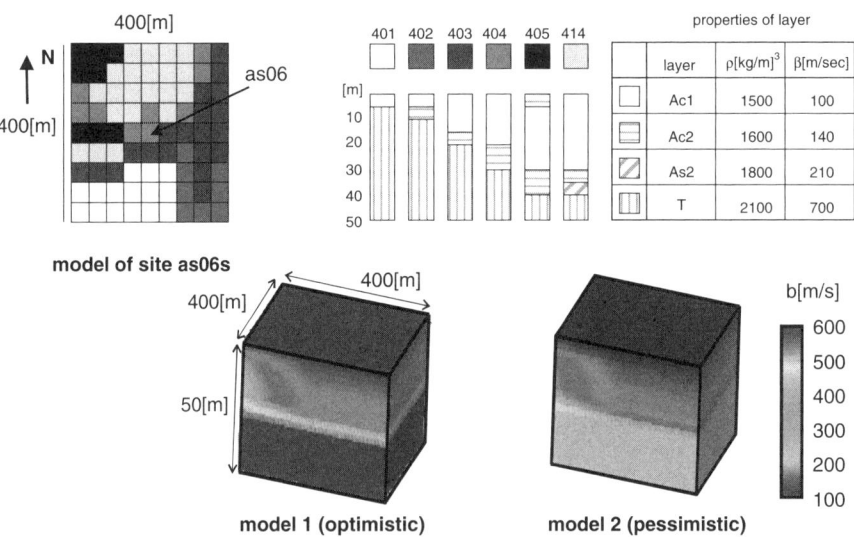

Fig. 11-8. Model for micro-analysis

the synthesized waves agree with the observed one fairly well up to 4 Hz. Also, in Fig. 11-11, we present the maximum velocity computed for the synthesized and observed waves. Among 12 seismograph sites, the maximum velocity observed at 10 sites are well reproduced by the numerical computation.

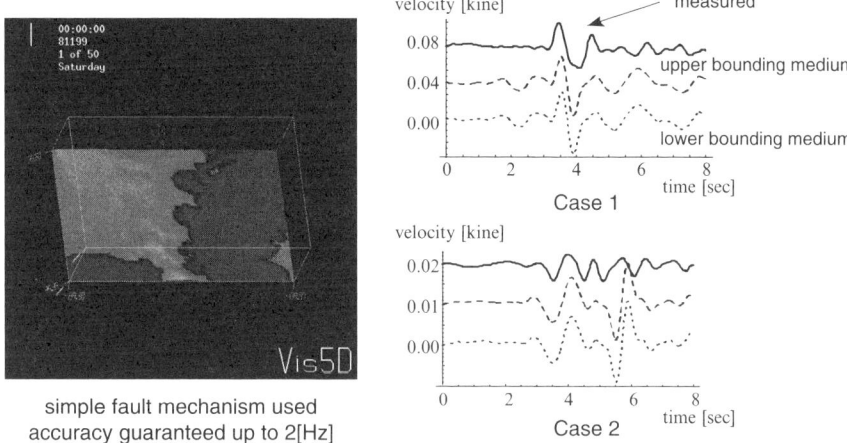

Fig. 11-9. Macro-analysis results

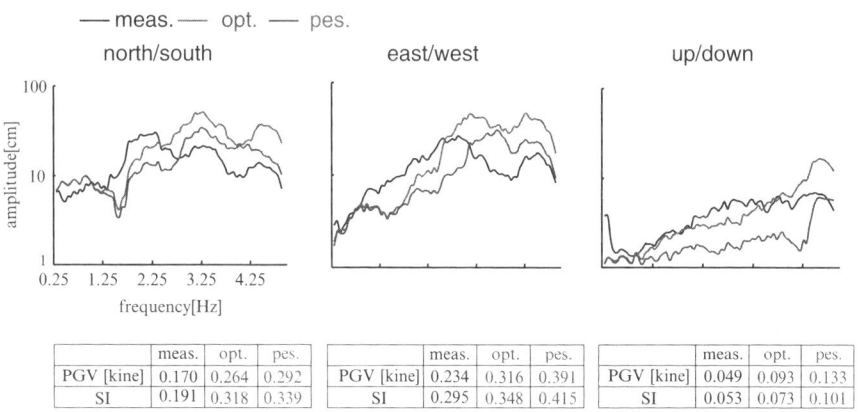

Fig. 11-10. Micro-analysis results (1)

11.3.4 Earthquake Hazard Prediction of IES

The earthquake hazard prediction of IES is the structure response simulation that is linked to the earthquake simulation. A computer model is constructed for a target town or a target city, with the help of the available GIS. For each building or structure, a structure model is constructed. Strong ground motion, computed at its site by using the ground-layer model, is input to the model, and the seismic response of the computer model is analyzed by using a suitable analysis method.

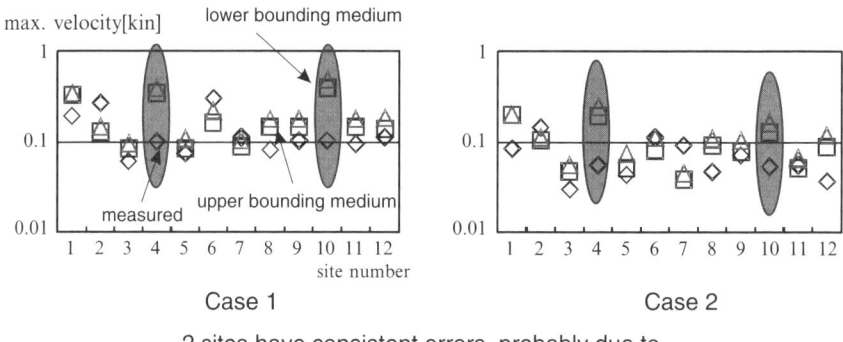

Case 1   Case 2

2 sites have consistent errors, probably due to
poor modeling

Fig. 11-11. Micro-analysis results (2)

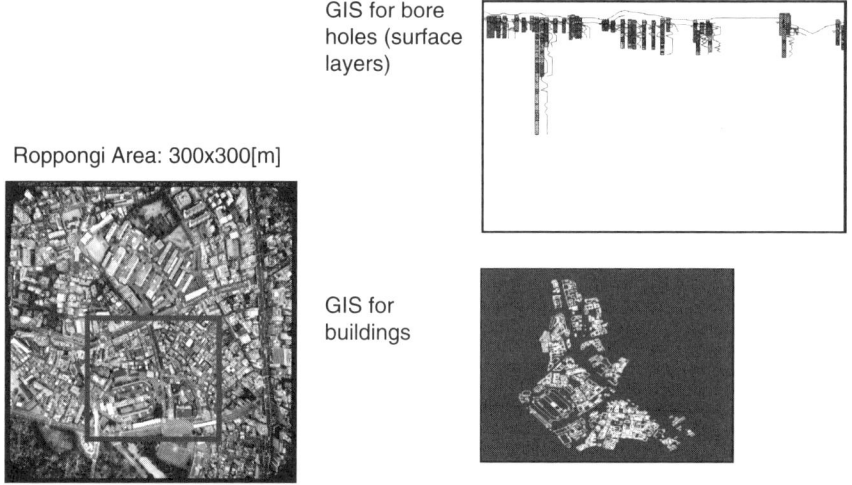

GIS for bore
holes (surface
layers)

Roppongi Area: 300x300[m]

GIS for
buildings

Fig. 11-12. Virtual town

As an example, we present a virtual town that is made for the old
Roppongi Campus of the University of Tokyo (Fig. 11-12). The area is 300
× 300 m, and GIS is used for boring hole data as well as for building configuration and property.

Here. a surface layer model and a set of structure models are constructed
(Fig. 11-13). The surfacelayer model consists of six distinct layers, which
have three-dimensional configuration and soil properties such as density,

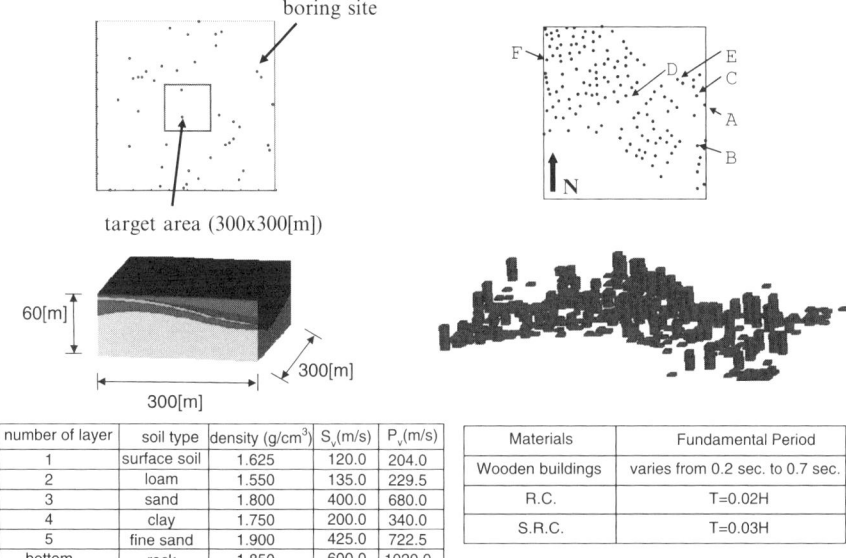

number of layer	soil type	density (g/cm³)	S_v(m/s)	P_v(m/s)
1	surface soil	1.625	120.0	204.0
2	loam	1.550	135.0	229.5
3	sand	1.800	400.0	680.0
4	clay	1.750	200.0	340.0
5	fine sand	1.900	425.0	722.5
bottom	rock	1.850	600.0	1020.0

Materials	Fundamental Period
Wooden buildings	varies from 0.2 sec. to 0.7 sec.
R.C.	T=0.02H
S.R.C.	T=0.03H

Fig. 11-13. Data and computer model

P- and S-wave velocities. A multi-degree-of-freedom model (MDOF model) is made for each of 147 residential buildings. The degree of freedom is the floor number, and the first and second natural modes are approximately determined in view of the building height and type; GIS provides information for building types, namely, wooden house, reinforced-concrete building (RC), and steel-reinforced building (SRC).

An earthquake hazard prediction is made by applying linear analysis of the MDOF model subjected to a synthesized strong ground motion. The results are visualized as a video clip that shows buildings shaking in the virtual town (Fig. 11-14). Every building has its own seismic responses, due to the strong ground motion, which is not uniform in the virtual town, and also due to the structural properties that are characterized as the MDOF model.

11.3.5 Development of Advanced IES

IES is a computer system that consists of many modules of numerical simulation programs (Fig. 11-15). Thus, an advanced IES is made by replacing or updating these programs. To easily plug a simulation program into IES, it has been designed as a federation-type data base. A particular interpreter

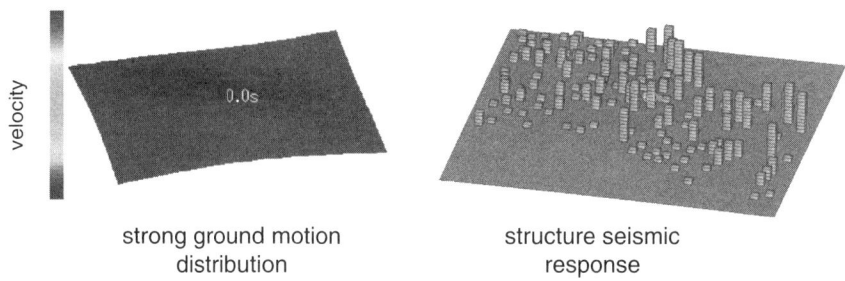

velocity

0.0s

strong ground motion
distribution

structure seismic
response

Fig. 11-14. Simulation of virtual town

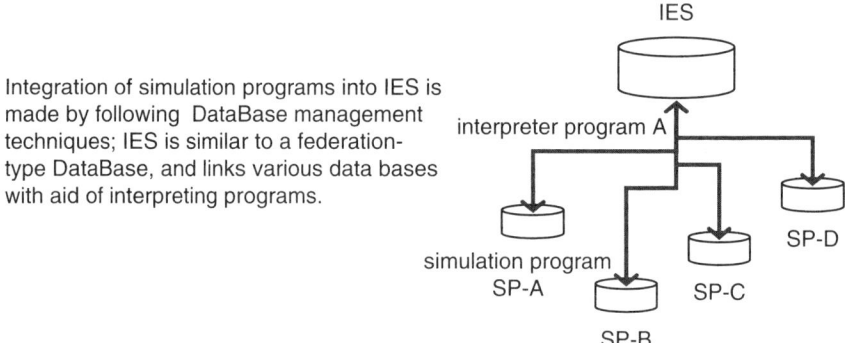

Integration of simulation programs into IES is
made by following DataBase management
techniques; IES is similar to a federation-
type DataBase, and links various data bases
with aid of interpreting programs.

IES

interpreter program A

simulation program
SP-A

SP-B

SP-C

SP-D

Fig. 11-15. Plug-in of numerical methods

program is made for each simulation program and this interpreter inputs
necessary data to the simulation program and gives back its output to IES,
converting data or results suitably.

An interpreter program is called a mediator in IES (Fig. 11-16). The basic
structure of the mediator is common to all such programs. This is because
the roles of the mediator are (1) to find a location of a target structure, (2)
to take data for it from GIS, (3) to select a site at which input strong ground
motion is computed, (4) to analyze input strong ground motion, (5) to put
all input data into the simulation program, (6) to take computation results
from the simulation program, and (7) to make visualization of the structure
response. A non-linear simulation program for structure seismic analysis is
plugged into IES by developing a mediator for the program without making
any change to it.

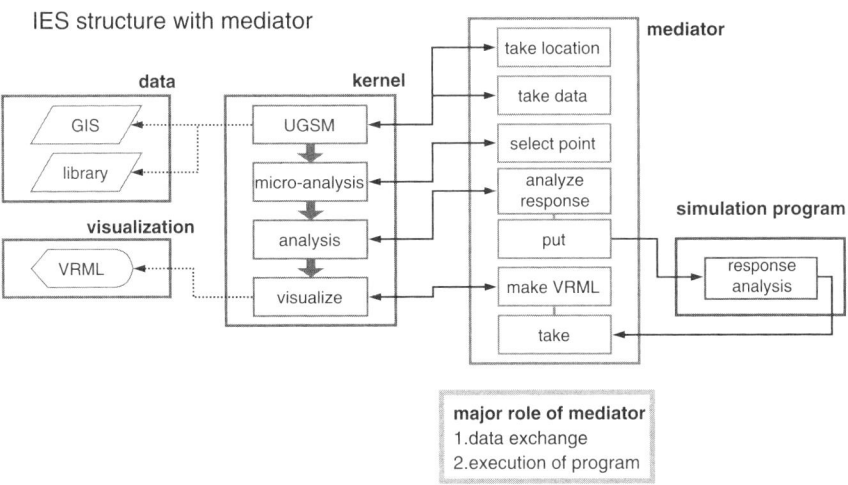

Fig. 11-16. Mediator as interpreting agent

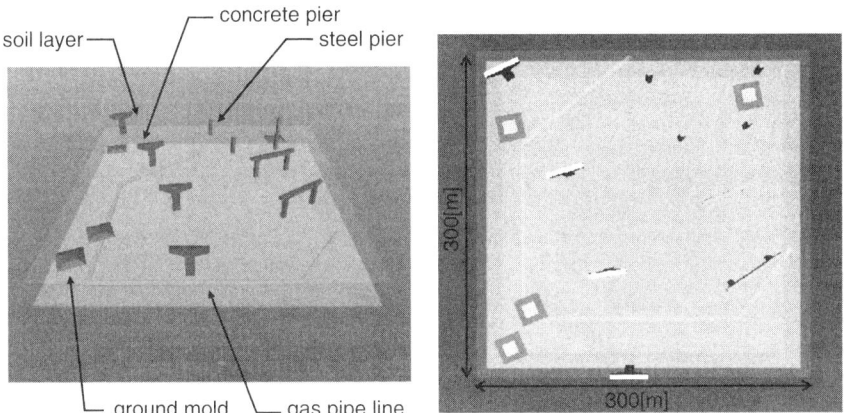

Fig. 11-17. Virtual city 3

We now present an example of an advanced IES, to which three non-linear simulation programs have been plugged in. The programs are for a concrete bridge pier, a steel bridge pier and a ground mold, and they have been made by professional researchers working with steel structures, concrete structures, and geotechnology, respectively. A computer model of virtual town is made consisting of four concrete bridge piers, four steel bridge piers and four ground molds, together with an underground gas-pipeline network (Fig. 11-17). A computer model is constructed for each structure in it.

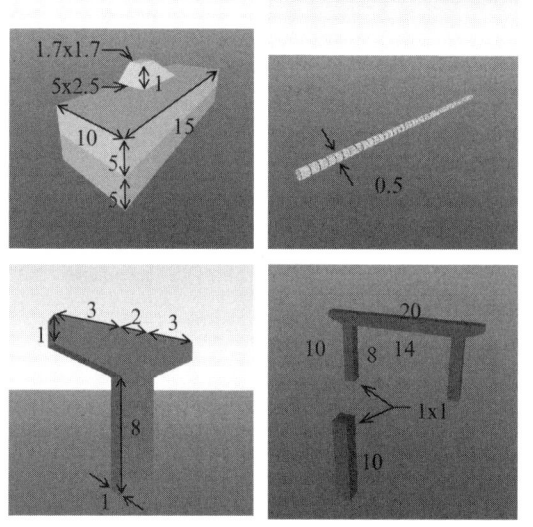

infrastructures
- gas pipe line
- concrete pier
- steel pier
- ground mold

particular material properties
& mechanism for
earthquake resistance

Fig. 11-18. Models of infrastructure in virtual city

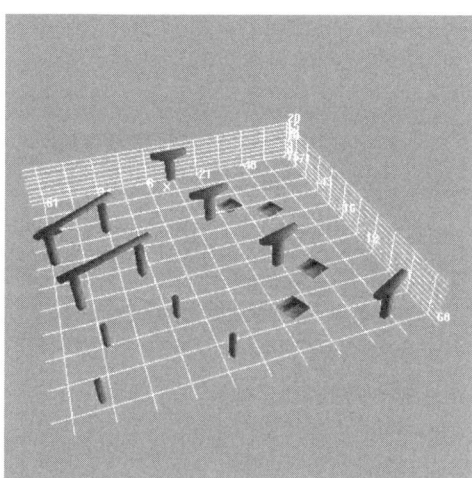

Fig. 11-19. Response of virtual city 3

For given earthquake waves, the distribution of strong ground motion is computed, and synthesized strong ground motion is input to each structure model. The non-linear seismic responses are then computed by using the plugged-in simulation programs (Fig. 11-19 and Fig. 11-20).

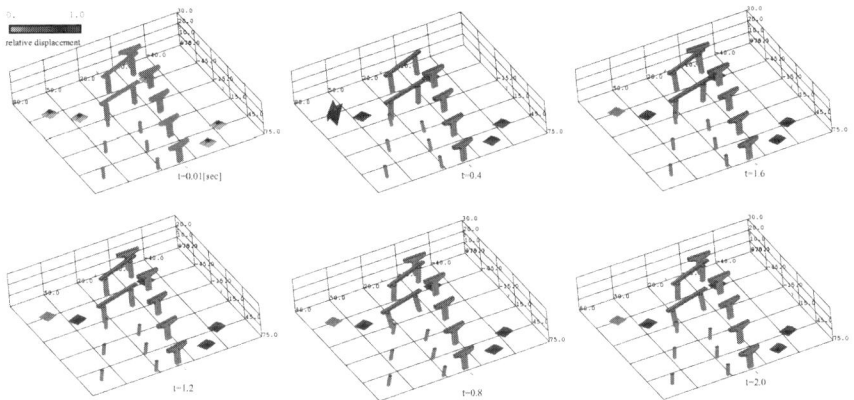

Fig. 11-20. Snapshot of response and damage

The reliability of the simulation program is high since they are based on earthquake-resistance design procedures and actually used for the design purpose. The failure or collapse of structures can be simulated if the input strong ground motion exceeds the seismic resistance of the structures.

The accuracy of earthquake disaster prediction is improved if an advanced simulation program that computes soil-structure interaction effects or interaction to near-by structures is plugged in. IES does not have to build such a program; it needs only a suitable mediator that interprets IES and the simulation program.

11.3.6 Action Simulation of Evacuation Process

As the first step of developing an action simulation, IES plugs in a multi-agent simulation that computes disordered evacuation process of a mass (Fig. 11-21). An agent has two data (thought and ability), and three functions (see, think and move). The data of thought is for movement and the data of ability is for physical ability; an agent sees the surroundings, thinks which way to go, and then moves toward that direction at a speed that the agent determines. Each agent has its own data so that a variety of human responses can be modelled. Randomly generating agents and distributing them in a model of damaged structures or in a network of narrow roads, IES makes a Monte Carlo simulation of a mass evacuation.

As an example, IES makes a model for a subway station that has three floors with four elevators. This model is used as a structure model with which seismic response of the station is computed for a given strong ground motion and an evacuation space model in which agents run up

◆ Utilization of *Intelligent* Agent

- ◆ input   internal variables (max. speed, intelligence, memory)
            external state

- ◆ output   see   surrounding environment, agent, etc.
            think   judge most suitable path
            move   go to next position

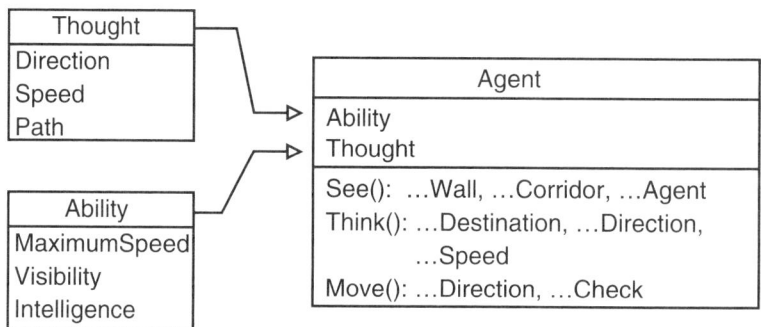

Fig. 11-21. Multi-agent simulation

- ◆ Evacuation Simulation
  - ◆ predict evacuation process in underground towns, high-rise buildings, department stores, schools, etc.
  - ◆ use intelligent agents to mimic people in panic state

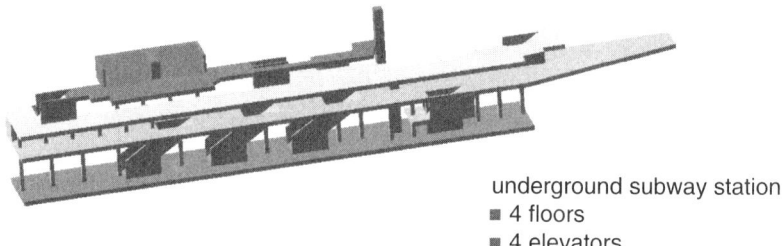

underground subway station
- ■ 4 floors
- ■ 4 elevators

Fig. 11-22. Action simulation: Evacuation simulation using MAS

to the upper floor (Fig. 11-22). The structure responses are computed by using linear FEM analysis; since the surrounding soil layers are included in the structure model, soil-structure interaction effects are computed in the simulation.

The evacuation process is simulated by generating two types of agents, quick and slow, so that the mixture of people moving at different speeds influence the mass evacuation. The process when the subway station is damaged by strong ground motion is compared with the process when the subway station is not damaged; the structure-response analysis of the subway station indicates that one stairway is damaged (it is assumed that a stairway is damaged when it reaches a critical stress).

The simulation shows that there is no significant increase in the evacuation process when 300 agents are distributed in the subway station and when one stairway is damaged by the strong ground motion. Also, it shows that the presence of slow agents do not influence the evacuation process drastically even though the evacuation time becomes longer as slow agents take longer time to move out from the model. These results are reasonable (Fig. 11-23). The agent simulation is used to make quantitative comparison for the safety of stations, by computing the evacuation time of the agents.

11.3.7 Integration Technique

In closing this section, we should point out that the key element of IES is the integration technique that combines GIS data with the asset of numeri-

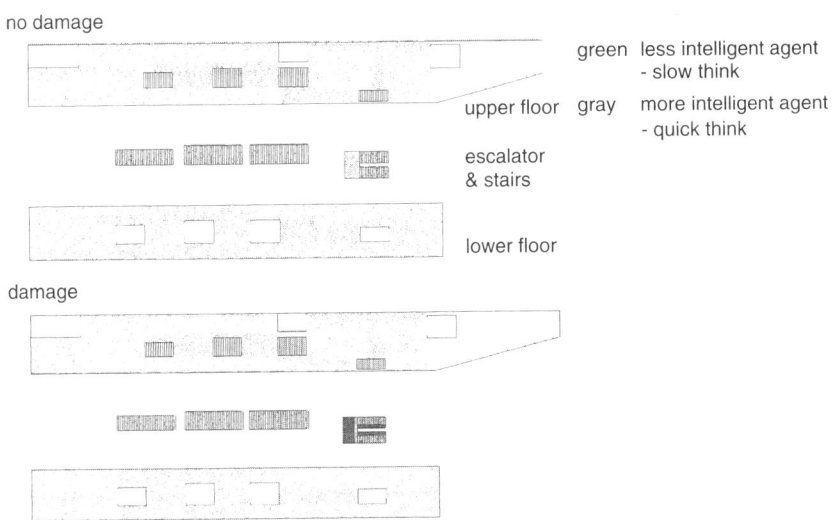

Fig. 11-23. Results of simulation (2)

cal analysis methods that have been developed for other purposes such as seismic design. In this sense, IES is understood as a platform for GIS and numerical analysis methods; in principle, any GIS and numerical analysis method can take part in this platform, and data and simulation are used to produce information for earthquake hazard and disaster for a given earthquake scenario.

Higher spatial resolution as well as higher reliability is needed for IES as a platform; latest GIS should be used and more advanced numerical analysis methods should be plugged in. Also, the integration technique should be improved. This is a new challenge for earthquake engineering because the integration technique has been studied in software engineering that is the foundation for developing larger, smarter, faster and more efficient programming.

11.4 Examples of IES Developed for Actual Cities

11.4.1 Kobe City

As an example of IES, we make a computer model for Kobe City, Japan, which was struck by the 1995 Hyogo-Ken Nanbu Earthquake (Fig. 11-24).

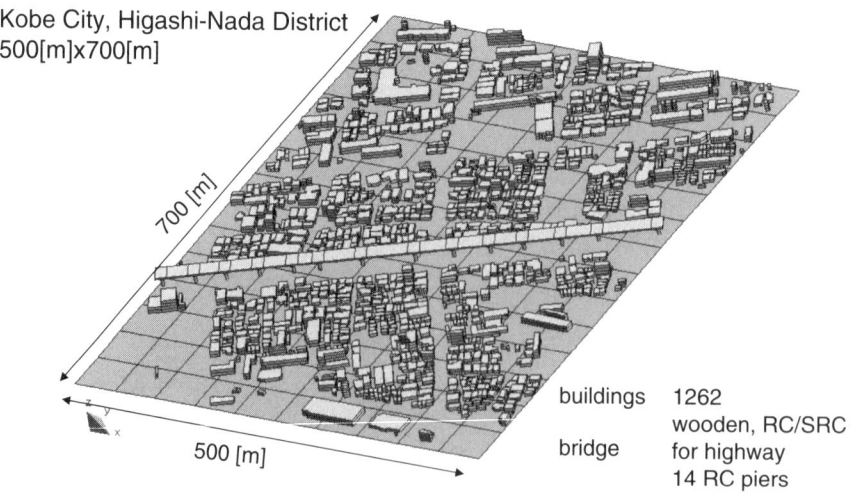

Fig. 11-24. Digital Kobe

The target is a 500 × 700 m domain of Higashi-Nada District; 500 m in the E-W direction and 700 m in the N-S direction. The surface ground layer model is constructed by using available boring hole data. There are 1,262 buildings (wooden houses, reinforced concrete buildings and steel-reinforced concrete buildings) and 14 reinforced concrete piers for Hanshin Expressway. MDOF models are constructed for all the buildings and used for the seismic response analysis; the first and second modes are made by assuming the natural frequency, damping coefficients and mode shape. An FEM model is constructed for the piers; the model is made from design blueprints.

We consider three scenarios of earthquakes (Fig. 11-25). Case 1 uses a measured waveform of the 1995 Hyogo-Ken Nanbu Earthquake (JR Takatori Station) to compute the strong ground motion distribution; the waves are input to the bottom of the surface ground layer model. Case 2 replaces the surface layer model with a simple stratified layer structure, excluding three-dimensional topographical effects of curved ground layers on the strong ground motion. Case 3 modifies the direction of inputting the waves from 90 degree to some extent.

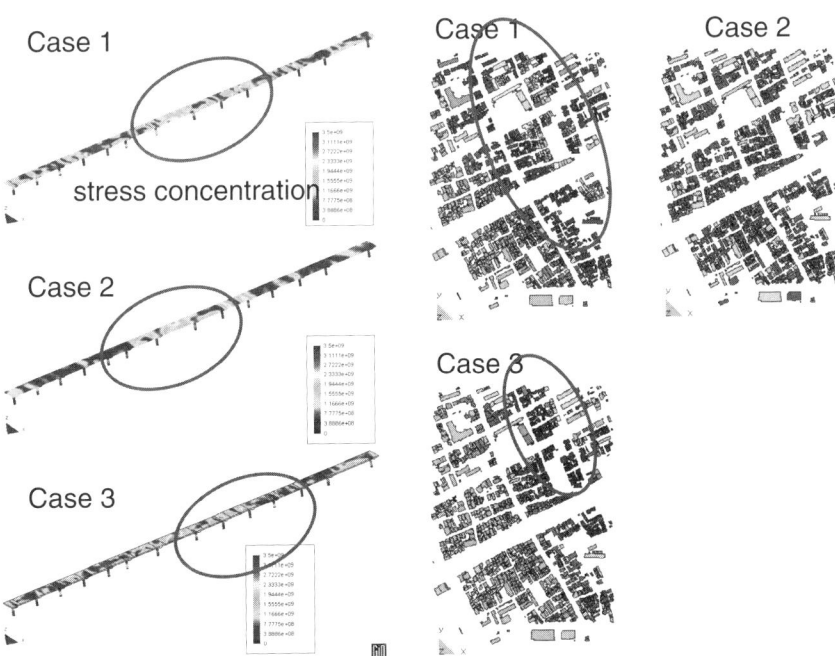

Fig. 11-25. Simulation results (2): Three earthquake scenarios

The resulting damage of structure responses are presented for the three cases. For the buildings, drift angle is used as an index of damage. It is shown that the concentration of damaged buildings changes depending on the earthquake scenario; Case 2 does not have damage concentration similar to Case 1, and Case 3 has a damage concentration the location of which is slightly moved from that of Case1. For the RC piers, the maximum stress is used as an index of damage. Like the buildings, the distribution and degree of damage change depending on the earthquake scenario. While the degree of damage for Case 2 is similar to that of Case 1, Case 3 has it to a larger degree.

The present example shows use of IES. It is well expected that earthquake disaster of Kobe City will be changed if another earthquake hits the city. The sequence of numerical simulations carried in IES gives quantitative information to the change in the disaster. Although overall damage, such as the number of damaged buildings or structures, may not be significantly different, the distribution of damaged buildings and structures is surely changed.

Since there is some uncertainty in predicting the source process for a possible earthquake, the prediction of earthquake hazard and disaster in a target city ought to have a certain range. IES could be a tool for estimating such a range.

11.4.2 Bunkyo City

The next example is Bunkyo City, Japan, where the University of Tokyo is located (Fig. 11-26). The key feature of this example is that the sequence of earthquake simulation, structure response simulation and action simulation

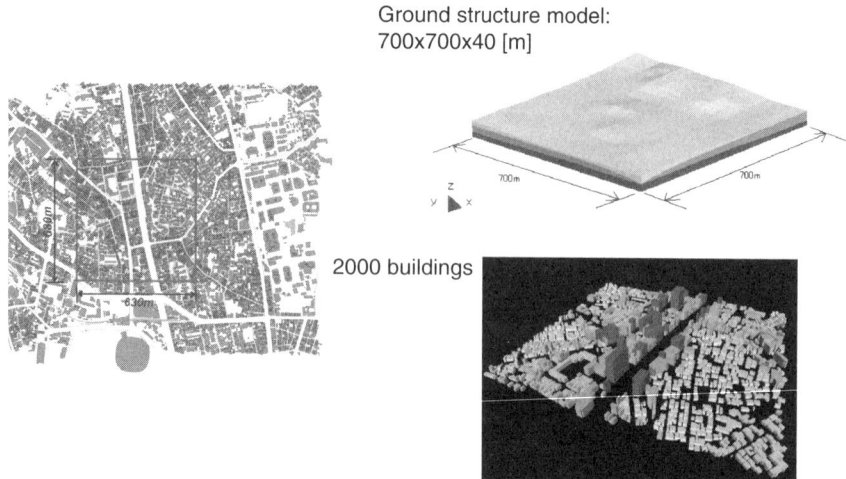

Fig. 11-26. Digital Bunkyo city

is carried out for a city model which is constructed by using a commercial GIS. This means that IES is applicable to other parts of Tokyo or Japan if GIS of the same kind is available.

Analyzed here is an underground structure model with a domain of 680×630 m; 680 m in the EW direction and 630 m in the SW direction. A ground surface layer model consisting of three distinct layers is constructed by using GIS which contains boring hole data. An MDOF model is constructed for each building by using a commercial GIS with detailed configuration data. GIS, which has data for structure information such as types or construction year, is not used to make these MDOF models. Building color indicates its height; red, green, or yellow is for a tall, medium, or short building, respectively.

The results of the earthquake simulation and the structure responses are presented. Like the previous cases, the distribution of strong ground-motion is not uniform due to the three-dimensional structure of the surface layer models, and some parts have larger concentration (Fig. 11-27). Building responses also vary depending on the input strong ground motion and the properties of the structure. An MDOF model constructed for a building is represented by a set of cubes; each cube is a mass that corresponds to a floor.

Using the same GIS as used for constructing MDOF models for buildings, we can make a road network model to execute the multi-agent simulation of the mass evacuation process; the domain to which the road net-work model is constructed is different from the one used for the earthquake and structure response simulations (Fig. 11-28). When some buildings are damaged or collapsed, we may not be able to use nearby roads. If a number of roads are closed, it surely influences the evacuation process. The multi-agent simulation is used to evaluate such influence of road destruction on the evacuation of the residents.

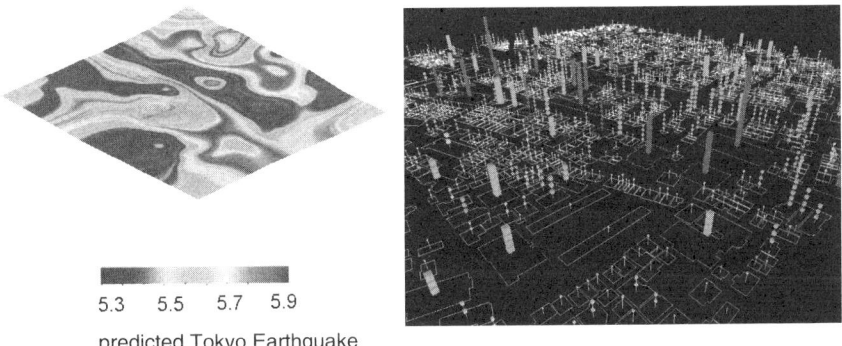

5.3   5.5   5.7   5.9

predicted Tokyo Earthquake

Fig. 11-27. Strong ground motion and structure response

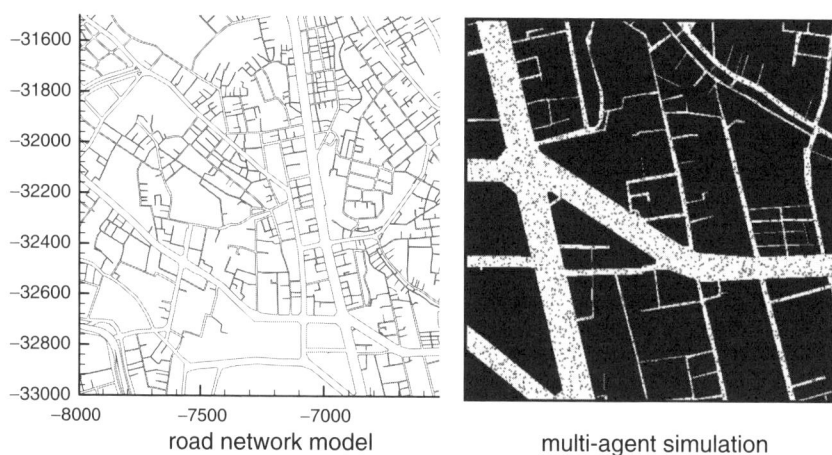

road network model   multi-agent simulation

Fig. 11-28. Evacuation simulation

11.5 Use of IES for Earthquake Disaster Recognition

The earthquake simulation of IES will be used as a next generation hazard map; it provides distribution of strong ground motion in finer spatial length scale that changes depending on a given earthquake scenario. The structure response simulation of IES provides information for damage of structures. Also, the action simulation for the mass evacuation process shows possible difficulties in escaping from damaged area. These predictions, however, have some errors due to the limitations of modelling. The underground structure model of the first earthquake simulation has some errors due to the limitation of geological survey and boring hole data. Structure models for buildings are constructed with some assumptions being made in determining the mechanical properties of the model. The accuracy of multi-agent simulation is limited since the agent is a simplified model of a man.

It is certainly true that IES is not a system that can make accurate predictions of earthquake hazard and disaster. The limitation of modelling underground structures, buildings or humans cannot be underestimated even if the most advanced numerical analysis methods are plugged in. The use of IES, however, is not in making accurate predictions. IES provides most reliable information for a possible earthquake disaster with highest spatial resolution, so that the first step of risk management—the recognition of risk itself—is improved. Common recognition of a possible earthquake hazard within a community is of primary importance to promote earthquake mitigation plans. IES provides the variability of a possible earthquake disaster that changes depending on an earthquake scenario and the community can understand a range of the earthquake disaster that can happen in the region.

References

Gruber, T.R. (1993) "A Translation Approach to Portable Ontology Specification", *Knowledge Acquisition* (5)2:199–220

Hammer, J., Gracia-Monlina, H., Ireland, K., Papakonstantinou, Y., Uhlman, J.D. and Widom, J. (1995) *Information Translation, Mediation, and Mosaic-Based Browsing in the TSIMMIS System*, in Exhibits Program of the Proceedings of the ACM SIGMOD International Conference on Management of Data, 483, San Jose, CA, June 1995

Hayashi, H. (2003) *Earthquake Mitigation for Safety of Life*. Tokyo: Iwanami.

Hirose, M., Tanikawa, T. and Endo, T. (1999) "A Taxonomy of Real and Virtual World Display Integration", in Ohta, Y. and Tamura, H. (eds) *Building a Virtual World from the Real World, in Mixed Reality—Merging Real and Virtual Worlds*. Berlin: Springer, pp. 183–197

Ichimura, T. and Hori, M. (2000) "Macro-Micro Analysis for Prediction of Strong Motion Distribution in Metropolis", Journal of Structural Mechanics and Earthquake Engineering, JSCE. 654/I–52: 51–61

Ichimura, T. and Hori, M. (2002) *Macro-Micro Analysis Method for Computation of Strong Motion Distribution with High Resolution and High Accuracy*. American Geophysical Union (AGU) Fall Meeting, San Francisco, CA, 6–10 December, 2002. S12B-1220

Kawano, H., Yamada, S., Kitamura, Y. and Takahashi, K. (2002) *Intelligent Information Technology on Internet—Information Search and Agent*. Tokyo: Denki University Press

Nagao, K. (2000) *Frontier of Agent Technology*. Tokyo: Kyoritsu

Nishio, S., Ohta, Y., Yokota, K., Nishida, T. and Satho, T. (1999) *Sharing and Integration of Information*. Tokyo: Iwanami

Osawa, Y., Minami, T. and Matsushima, Y. (1999) "Dynamic Analysis for Structures", *Shokokusha*, 38

Shiono, K., Wadatsumi, K. and Masumoto, S. (1987) "Numerical Determination of the Optimal Bedding Plane", *Geoinformatics*, 12: 299–328

Uhlman, J.D. (1997) "Information Integration Using Logical Views", *Proceedings of ICD'97. Springer LNCS* (1186): 19–40

Yagawa, G. and Kanto, Y. (1999) *Introduction to Object-Oriented Computational Mechanics*. Tokyo: Baifukan

Yamanaka, H., Sato, H., Kurita, K. and Seo, K. (1999) "Array Measurements of Long-Period Microtremors in Southwestern Kanto Plain, Japan", *Zisin*, 51:355–365

12. Risk Assessment, Management and Monitoring of Infrastructure Systems

Yozo Fujino and Tsuyoshi Takada

12.1 Introduction

Urban infrastructure constitutes physical facilities such as road and highway networks, buildings, railways, dams, bridges, underground areas, water supplies and so on, all of which provide essential services to support and sustain civilian life in an urban area. It is a public asset that is accessible to everybody and serves as a basis for their economic and social activities. Therefore the economy of one nation is heavily dependent upon the satisfactory performance of its infrastructures. In this paper the concept of risk assessment and management of urban infrastructures are discussed. Focus of discussion is put on the risk due to natural disaster and infrastructure deterioration, and on the sustainability of infrastructure. The concept of risk assessment is herein presented first, followed by examples of risk assessment and management due to natural disasters. The last part introduces infrastructure monitoring as an essential tool for risk reduction. Current practices in monitoring of infrastructure system, such as bridges and railways, are demonstrated.

12.2 Concept of Risk Management

12.2.1 Definition of Risk

The word "risk" originates from an ancient Italian word *risicare*, which means "to dare." It has a positive connotation, in the sense that risk is perceived as a choice rather than a fate (Bernstein (1996)). This positive meaning resonates in the modern concept of risk management. Many people consider that risk management is a new concept, while in fact it is an old one. There are some definitions of risk that are commonly perceived. The international standard organization (ISO/IEC (2002)) defines risk as the combination of probability of an event and its consequence. More specifically, risk is defined as the quantity of probability of damaging events multiplied by the loss due to that damage. To this end, it is important to quantify the probability of damage and to estimate its consequence in term of monetary units. From a psychological viewpoint, a rather similar definition is given (Hirota (2002)). In the concept, risk is defined as a product of the magnitude of loss and probability of having the loss. The magnitude of loss depends on the event or sources that threaten human lives or health such as accidents, hazards, and peril.

In this paper we define risk as a measure of danger that undesired events represent for humans, the environment, or economic values. Later on in this paper a specific type of risk regarding structures will be defined. In general the risk is a combination of likelihood of an event and its consequence (Takada (2006)). The consequence can be limited for one event or, in some cases, related to several events. It can be positive or negative, whereas in the case of safety aspects it is always negative. The consequence is expressed qualitatively or quantitatively. Based on the definition above, risk can be formulated as:

$$R = P \times C \tag{12-1}$$

where:

  R = risk, quantified in terms of monetary unit or number of casualty

  P = probability of occurrence of the event

  C = consequence or loss of event

The probability of occurrence refers to how often a specific scale of event occurs in a specified lifetime, whereas the consequence quantifies the results of that event as loss in term of monetary loss or fatality. Note that Equation (12-1) includes uncertainty as indicated by the probabilistic term. This would sometimes make it difficult to explain since most people tend to think in a deterministic way.

12.2.2 Organization of Risk Management

According to the ISO document, risk management can be divided into two main parts, namely, the risk assessment and control (Fig. 12-1). The assessment part consists of risk analysis and evaluation, while the control part deals with decision making and monitoring. In the risk analysis the

Fig. 12-1. Risk assessment in the framework of risk management (After ISO/IEC (2002))

first essential step is hazard identification. In this process we identify all set of circumstances that have potential for causing events with undesirable consequences. The second step is risk estimation, where the potential damages caused by the hazards are quantitatively estimated. Risk evaluation consists of two steps: one is risk acceptance, in which the criteria for accepting the risk are determined; and the other is risk treatment, where we identify the range of options for treating risk, assessing those options, and preparing risk treatment plans and implementing them. The effectiveness of risk control relies on the way we communicate with the people involved. It is a difficult task, since the majority of people do not bear in mind and even not try to think about the likelihood of accident or hazards.

The first step in risk assessment is to define the goals of risk management, which are implemented in a clear context. For instance, in case of the existing civil infrastructures, one important safety aspect is seismic safety. Therefore in this context, the goal of risk assessment is to investigate performance of the structures against earthquakes by investigating their present condition. Another example is the risk assessment conducted by a government before constructing a chemical plant. In this case, the government has the responsibility to convince people living in the vicinity that the plant structure is safe. Before communicating the risk with the people, a comprehensive risk analysis must be performed. This includes: (1) identification of potential hazards related to the structure, (2) defining and estimating the consequences of the hazards, and (3) providing countermeasure schemes to reduce the risk.

12.2.3 Risk Classification, Quantification and Comparison

Risk can be classified into four categories according to the frequency of its occurrence and consequence (Lewis (1990)). The first category is the risk with the well-known probability of occurrence and consequences, such as fire and traffic accidents. This type of risk is relatively easy to assess, since past data are readily available and the consequences can be easily estimated statistically. The second category is the risk that rarely occurs but is inevitable, and has significant consequence. Big earthquakes and meteorite impacts are the events that fall into this category. The third category is the unprecedented risk that has even more significant consequences, such as nuclear war or global contamination. Risk of type two and three are very difficult to assess. The absence of such events in the past makes the prediction of its probability of occurrence difficult, let alone any estimation of the consequence. For the unprecedented type of risk, predictions using mathematical models, mechanics, and social science are the only approaches in defining and quantifying the consequence. The fourth category is the undetectable risk, such as the effect of chemical material and low level radiation. Quantification of the last type of risk is even more difficult due to its invisible character.

It should be borne in mind that despite the probabilistic definition of the term, risk must exist objectively (Hayashi (2000)). This implies that the risk should be scientifically evaluated using technically sound procedures and acceptable indicators. The scientific nature of risk assessment implies that the evaluation process should be objective, transparent, and performed by the experts in that field. Moreover, when an assessed risk is publicized with its reasoning, it should be discussed simply scientifically. An interesting example on this is the ongoing public discussion on the risk of nuclear power plants. It is common that those who are against the nuclear power plant oppose it not for the risk reasons but simply because they reject the idea of nuclear. This resistance sometimes draws them into nonscientific arguments and distorts public opinion.

The risk curve is often used to quantify various type of risk. This curve quantifies the risk in terms of probability of occurrence and the expected consequence (Fig. 12-2). The curve is also utilized to compare the risk of one event with the other events. In an example provide by USNRC (1981) and (Stewart and Melchers 1997) for instance, it is shown that the risk of nuclear power plant failure is much lower than the risk of traffic accidents or any other man-made hazardous events.

Once estimated, the risk can be evaluated according to the previously assigned acceptable levels. To this end, the concept such as ALARA/ ALARP (As Low As Reasonably Attainable/Possible) is adopted in order to determine the acceptability and to decide whether the risk should be treated

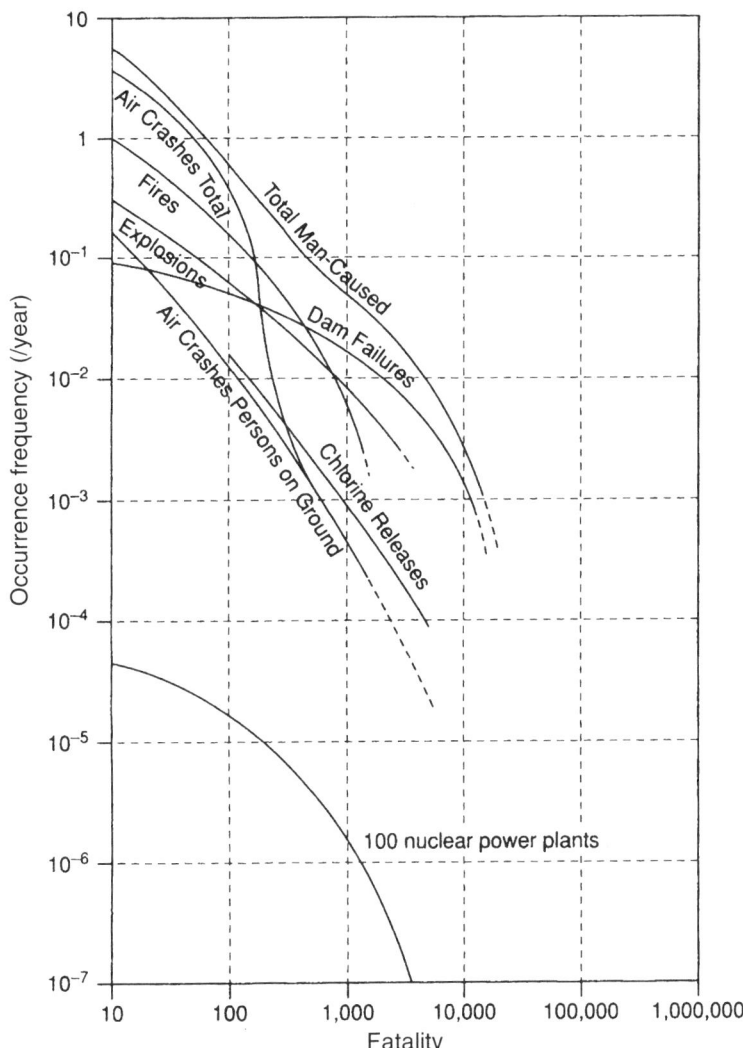

Fig. 12-2. Risk curve and comparison (After USNRC (1975))

or accepted. The ALARA/ALARP describes a range between two possible extremes: the acceptable risk at current levels and the unacceptable range. When a risk assessment indicates that the risk is acceptable at the current level, countermeasures should be taken to keep it at that level. However, if the risk is at an unacceptable level, countermeasures should be taken to reduce the risk. At the intermediary range, the risk is accepted only on the ground that: (1) its reduction is unrealistic, or (2) the cost of its reduction is grossly disproportionate to the benefits gained.

12.3 Risk Assessment in Earthquake Engineering

In this section, we implement the risk assessment to evaluate the seismic risk of a building. The following factors are considered in formulating the seismic risk.

- Intensity of input ground motion. A quantity such as Peak Ground Acceleration (PGA) is commonly utilized as an indicator of the severity of earthquake ground motion input.
- Site amplification. The ground motion must be modified to take into account the geologic and stratigraphic condition pertaining to the site. The site amplification factor measures how large the ground motion from engineering bedrock level is amplified before it reaches the surface level. When a building is located on a soft soil layer, the risk of failure during earthquake increases, since the seismic wave is greatly amplified.
- Estimation of building strength against earthquake. Using the input of ground motion at the surface level and by estimating the strength of a building via fragility data, we can construct the probability of failure. The probability of failure of the building is utilized to estimate the consequence in term of monetary loss or fatality.

There are four damage states of a building according to the level of horizontal deformation. They are ranked in proportion to the severity of damage as: (1) no damage, (2) partially damaged, (3) half collapsed, and (4) totally collapsed. Once the damage state is defined, we can estimate the probability of occurrence and the subsequent consequences.

12.3.1 Seismic Hazard Map

As mentioned above, the seismic risk assessment can be divided into two main parts: seismic hazard assessment and structure vulnerability assessment. One approach toward estimating the level of seismic hazard is by using the seismic hazard curve. This curve shows the relationship between seismic intensity such as the Peak Ground Acceleration (PGA) and its exceedance probability over a year. The seismic hazard curve shows the likelihood of a particular level of earthquake shaking that can be expected at one place during a given period of time.

In principle, each site has its own seismic hazard curve. Therefore curves for nearby sites can be correlated to present seismic hazard in a more comprehensive way in the form of a probabilistic seismic hazard map.

The seismic hazard maps typically depict the seismic intensity of a particular area under the condition of the probability exceedance. For example, Fig. 12-3(a) shows the ground motion intensity according to the Japan Meteorology Agency (JMA) with a 5 percent probability of exceedance in 50 years. Fig. 12-3(b) shows the ground motion intensity of United States in terms of peak ground acceleration (percentage of gravity) with a 10 percent probability of exceedance in 50 years.

The seismic hazard maps estimate the relative hazards considering both the severity of the potential ground motions and the frequency of occurrence. They also provide a way to compare hazards of different locations quantitatively. The maps usage includes: (1) educating the public about the earthquake hazards, (2) engineering applications and seismic zonation for building codes, and (3) hazard response planning.

12.3.2 Fragility Analysis of Buildings

The equally important aspect of seismic risk assessment is the identification of structural seismic vulnerability associated with various states of damage. One widely accepted practice to present the vulnerability information is in the form of a fragility curve. The fragility curve represents the probability that structural damages under various levels of seismic excitation exceed the specified damage states. There are two types of fragility curves: (1) the empirical fragility curves, which are developed using data of structural damages observed from past earthquake events, and (2) the analytical fragility curves, constructed by numerically simulating seismic response via structural dynamic analysis.

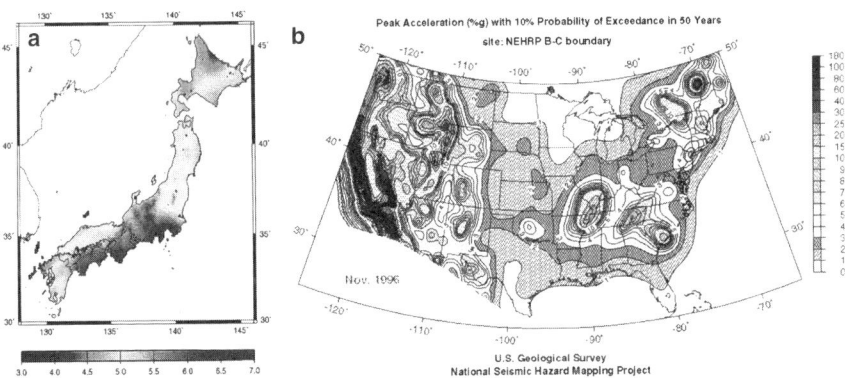

Fig. 12-3. (a) Seismic hazard map of Japan. **(b)** Seismic hazard map of United States

When data of structural damages from earthquake events are available, the empirical fragility curves are more beneficial than the analytical ones. For instance, such curves were constructed on the basis of structural damages observed during the 1995 Kobe earthquake. The data for structural damage was sorted according to: construction year, structural type (e.g. wood, reinforced concrete, steel, and light-gauge steel), number of floors, total floor area, and its basic function (e.g., condominium, individual house, commercial use and factory). Each curve is drawn for a specified damage state. For instance, moderate damage (the state when significant structural damage is visible and permanent deformation remains) and severe damage—when large portions of structure are damaged and the permanent deformation may induce the collapse.

Fragility curves are drawn with respect to the ground motion intensity such as Peak Ground Acceleration (PGA) or Peak Ground Velocity (PGV). For the corresponding excitation level, the damage ratio is computed often by assuming a lognormal distribution. For the cumulative probability $P_R(\bullet)$ of occurrence of the damage equal or higher than rank R is given as

$$P_R(PGV) = \Phi[(\ln PGV - \lambda) / \zeta]   \tag{12-2}$$

where $\Phi[x]$ denotes the cumulative distribution function of the standard normal distribution, while λ and ζ are the mean and standard deviation of *ln* PGV respectively, that are obtained from the type of structure and the construction year.

Figure 12-4 shows an example of the fragility curves for severe damage case computed based on the 1995 Kobe earthquake. The comparisons of probability of damage for three types of construction material are depicted in Fig. 12-4(a). In Fig. 12-4(b) the probability of damage of reinforced concrete structures is presented according to the construction year. Using a typical fragility curve one can predict the likelihood of certain damage level for a given ground motion intensity. For instance in Fig. 12-4(c), the fragility curves show that the likelihood of the 40 cm/s PGV causing severe damage and moderate damage to a wood structure constructed after 1981 is 0.5 percent and 8.1 percent, respectively.

12.3.3 Comparison of Seismic Hazard with Other Natural Hazards

Risk of other natural hazards such as wind and snow damage can also be assessed using a similar approach. In the case of wind hazard, the maximum wind velocity is crucial for determination of wind hazard, whereas for the snow hazard, the snow depth or snowfall rate is used as the hazard

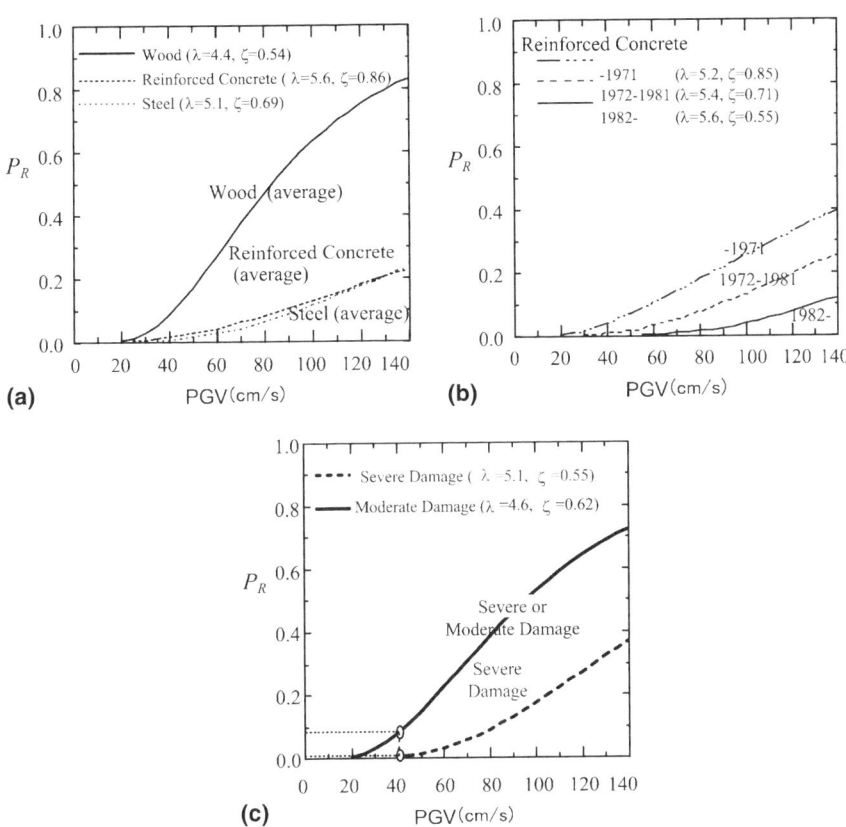

Fig. 12-4. (**a**) Fragility curves computed based on 1995 Kobe earthquake. (**b**) Fragility curve of reinforced concrete structure. (**c**) Fragility curve of wood structure constructed after 1981

indicator. Similar to the seismic risk assessment, the structure vulnerability due to wind and snowfall can be computed based on the empirical data or numerical simulations in the form of wind-induced fragility and snowfall fragility curves respectively. The risk due to the wind and snow are then calculated by combining the probability of failure of a specified structure for a given intensity of wind velocity and snowfall rate. The consequences of such events are quantified in terms of economical losses and casualties.

A comprehensive study of natural hazards and the risks they present can lead to an overall natural hazard risk assessment of an interested area. Takada (2006) presented such a study involving 10 big cities in Japan. In this study, the hazard map is based on the Architectural Institute of Japan

(AIJ) Load Recommendation (2004) consisting of three hazard maps with a return period of 500 and 100 years: namely, the seismic hazard map in terms of PGA on the engineering bedrock, the wind hazard map in term of wind velocity, and the snowfall hazard map in term of maximum snow depth.

The vulnerabilities of structure due to the three natural hazards are defined using the corresponding fragility curves. Empirical fragility curves are constructed on the basis of the observed fragility data with three damage states (partially, half and totally collapsed) using the following formulations:

- Seismic fragility estimation (Murao and Yamazaki (2000)):

$$P_{Ei}(T) = \Phi\left(\frac{\ln v(T) - \lambda_{Ei} - \ln Z}{\zeta_{Ei}}\right) \qquad (12\text{-}3)$$

- Wind fragility estimation (Kondo et al. (2002)):

$$P_{Wi}(T) = \Phi\left(\frac{\ln W(T) - \lambda_{Wi}}{\zeta_{Wi}}\right) \qquad (12\text{-}4)$$

- Snow fragility estimation (NIED (1982)):

$$P_{Si}(T) = \Phi\left(\frac{\ln d(T) - \lambda_{Si}}{\zeta_{Si}}\right) \qquad (12\text{-}5)$$

Using information derived from the seismic, the wind, and the snow hazard maps as well as their corresponding structural fragility estimations, the overall risk due to natural hazards are herein defined. Figure 12-5 shows an example of risk estimation for an individual house computed in 10 cities in Japan (Takada (2006)). From this figure one can notice that earthquakes are the greatest risk for individual houses in most cities in Japan. Shizuoka city in particular has the largest earthquake risk. The wind risk comes second in most cities, except for Tokyo, Sapporo and Fukuoka, where it comes first. Note that earthquake risk varies and depends on the region where the cities are located, while the wind risk is more evenly distributed. Also note that of all cities observed, snow does not pose significant risk as compared to both seismic and wind.

From hazard maps we can proceed to the risk maps, where the overall potential hazards are quantified. Such comprehensive overall risk maps are very useful to: (1) educate the public about the natural hazards in their region, (2) provide comparison of risk that each natural hazard poses in different areas and different types of structure, (3) assist the decision maker on the risk reduction countermeasures as well as budget allocation.

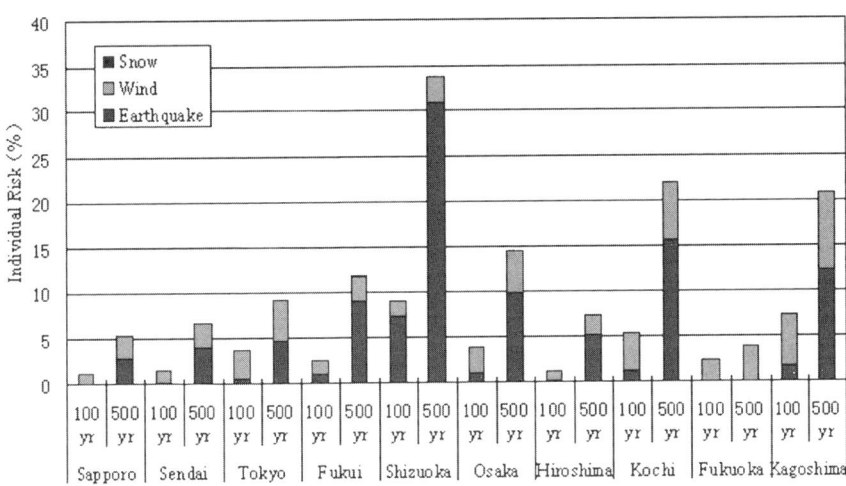

Fig. 12-5. Comparison of individual house risk at ten big cities in Japan (After Takada (2006))

12.4   Urban Infrastructure: Its Characteristics and Challenges

The amount of urban infrastructure has accumulated during years of development and inevitably created some difficulties regarding management and sustainability such as:

1. Financial responsibility

Generally people realize that they are responsible for subsidizing infrastructure development and management expenditures through the tax system. However, it is often uncertain how this expenditure should be dispersed. Concerning the temporal responsibility, the question raised is: when should the cost be paid? Should it be paid now by the current generation that would not yet enjoy the benefit, or later by the future generation who would enjoy the benefit? Or should it be distributed over time? Spatial-wise the question is: who should be paying for the infrastructure expenditure? Should the people who live near the facilities and thus directly benefit, be responsible? Or should people who live far away from the site also share the burden?

2. Scatter in temporal and spatial domain

Urban infrastructures are physically scattered in time and in space, lying within two extreme boundaries: the macroscale extreme and the microscale extreme. When it comes to the macroscale, such as the global scale, the problem is usually approached and solved statistically. On the other hand, for the microscale problem, nanoscale or micro mechanics modeling would be more suitable. The urban infrastructure scale is somewhere in between—commonly referred to as mesoscale. It is subjected to many uncertainties depending upon their conditions, locations, type of loadings and usages.

3. Social and psychological consequence of infrastructure deterioration
During the service life, physical facilities are subjected to inevitable deterioration due to material degradation and aging. Also, they might be subject to extreme events that cause severe deterioration or an unexpected accident. Direct consequences of such events may hamper social and economic activities. In addition there are also sometimes-unforeseen psychological consequences. When one accident involving specific infrastructure occurs, such as the failure of a bridge, people become very concerned about the safety of the same facilities at other places. People tend to perceive a physical facility not as unique entity with its unique problems, but rather as the general entity with identical problem, so that the failure of one facility may create a public anxiety and lead to the question of the safety of the typical facilities at different places.

12.4.1 Impact of Natural Disasters on Infrastructure

Experience has shown that modern infrastructures are always subjected to various types of disasters caused either by natural, accidental, or deliberate hazards. The largest portion of infrastructure loss is due to natural disaster. In Japan for instance, the loss suffered due to natural disaster from 1970 to 2004 is approximately US$11 billion, which comes second in the world after the United States. Of the world's total infrastructure loss caused by natural disaster, Japan's share is 15 percent, which is one-sixth of the total loss in the world (Fig. 12-6(a)). From the continental viewpoint, almost half of the losses take place in Asia. The amount of loss in Japan is particularly high—half of the total loss suffered by all Asian countries combined.

Looking at history, Japan has suffered tremendous loss in terms of property and human causalities as a result of natural disasters. Fig. 12-6(b) shows that in the past few decades the property loss has generally increased, but the human loss has significantly decreased. The increase in property loss is somehow reasonable considering that development acceleration in recent decades have resulted in accumulation of infrastructure stock. The decrease in human loss indicates that technologies and investments are effective in increasing human safety during natural disasters.

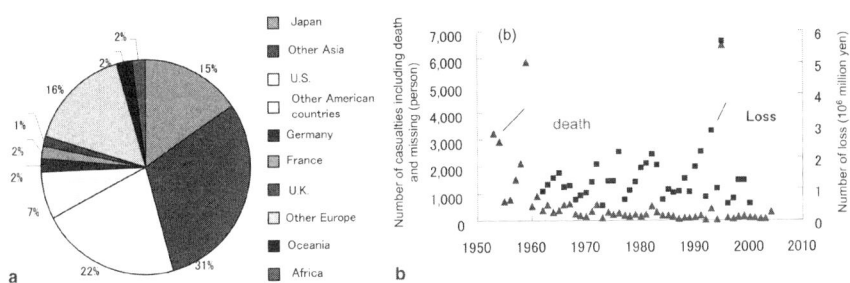

Fig. 12-6. (a) Casualties and loss due to natural disaster worldwide. **(b)** Casualties and loss due to natural disaster in Japan

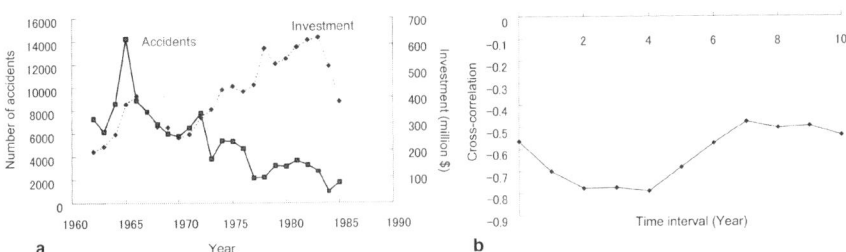

Fig. 12-7. (a) Comparison of number of accident and investment made by Japan Railways. **(b)** Cross-correlation between number of accident and investment

12.4.2 Rationalization of Disaster Prevention Investment

In order to prevent natural disasters and their related casualties, Japan has allocated large amount of resources for natural-disaster-related activities such as research, prevention, maintenance, and recovery. On the average, Japan spends from ¥300 to ¥400 billion annually for these activities. Studies show that investment in disaster prevention has increased the safety of infrastructure system. Figure 12-7(a) for instance, demonstrates that the increasing number of investment made by Japan Railways has improved the performance of the railway system significantly and therefore reduced the number of train accidents.

Figure 12-7(b) shows the cross correlation between the investment and the number of accidents. The negative correlation in this figure implies that the number of accidents will not decrease unless accident prevention investments are made. The rather flat correlation suggests that the effects of disaster prevention investment do not immediately materialize. They lag after some time, but remain for a long time. On the contrary, the effects of disaster prevention investment in engineering plants appear immediately, but do not last for a long time. In other words, if we do not allocate sufficient budgets for infrastructure maintenance, we may not experience the immediate negative effects. In the long run, however, we may suffer from greater consequences. These results emphasize the slow impact of infrastructure investment and the long-term management strategy it requires.

12.4.3 Challenges in Japan's Infrastructure Stock

Japan's infrastructure stock has accumulated significantly in the past few decades owing to the large increase in its Gross Domestic Product (GDP). Compared to the United States, however, the infrastructure development in Japan lags a few decades behind. A good example of this comparison is the development of bridges. In the United States, development of bridges hit the highest point twice between the 1930s and 1960s, while in Japan it reached the peak around the 1970s (Fig. 12-8(a)). This means that a large amount of bridges in US have been used for more than 50 years, and the majority of them are now aging. During the mid 1970s, deterioration and aging of bridges were serious problems in the US. Many accidents, such as bridge malfunction and collapse, occurred during this period. To resolve the problem, a bridge management system was set up. It was a nationwide mandatory program that mandated bridge assessment through visual inspection and performance evaluation in the form of a rating system. The assessment results and inventory data were made accessible to the public and have alerted people of how serious the problems were. Feedback from the assessment had prompted the US government and Congress to take immediate action and allocate more budgeted funds for bridge maintenance.

Many Japanese bridge experts consider that the bridge condition in Japan now is generally not as bad as the condition was in the US at that time. Nevertheless, as time passes, evidence shows that more and more bridges in

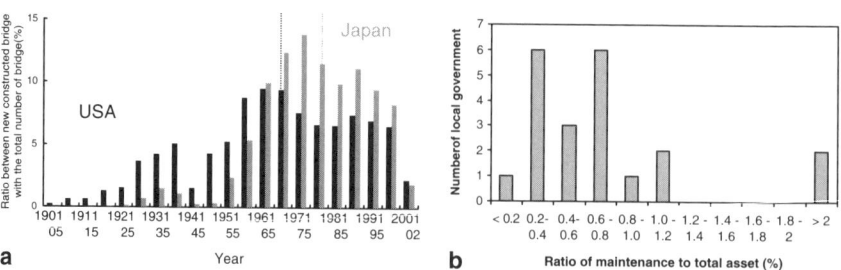

Fig. 12-8. (**a**) Comparison of bridge construction period between us and Japan. (**b**) Ratio of maintenance and total assets allocated by local government in Japan

Japan are aging and deteriorating. Consequently, it is anticipated that Japan will soon face the same problem.

Visual inspection alone, however, is often insufficient. Some recent bridge accidents have shown that the bridges were in fact visually inspected prior to the accident but no damage or failure symptoms were apparent. The lack of details and objectivity in visual inspection can be made up with a more comprehensive monitoring system using quantitative inspection. This was initiated in the US in a form of a long-term bridge performance monitoring program that included continuous monitoring for over 20 years. The program consists of quantitative inspection, bridge monitoring, and in some cases forensic studies of the soon-to-be demolished bridges. With this approach, the degradation level as well as the prediction of the bridge's remaining service life can be quantified more accurately. This will lead to an efficient infrastructure stock management.

Infrastructure maintenance covers much a wider range of areas beyond bridge monitoring. Infrastructures such as school buildings, for instance, pose a serious maintenance challenge. A large number of university buildings in Japan are now aging, while the annual budgets for maintenance and retrofit are very limited. There are over 100 national universities in Japan. If each university has 10 buildings in need of retrofit, where each retrofit requires US$5 million, the sum of total annual budget required nationwide would reach US$5 billion, while the total allocated budged is only US$700 million, including the construction of new facilities. There are around 60,000 elementary and high school buildings nationwide with even more severe conditions. The total costs of having them retrofitted would exceed US$60 billion, while the total allocated budged nationwide is only US$550 million, including the construction of new facilities.

The national road system also has a similar problem. The central government is responsible for seismic retrofit on 4,500 locations, while local governments cover around 36,000 locations. The total amount of budget required for seismic retrofit of this road network is estimated to be around ¥4,050 billion yen. To set up a natural disaster mitigation and prevention system, the central government is responsible for 3,000 locations, while the local governments are responsible for 68,000 locations. Therefore, supposing that one natural disaster and prevention program requires ¥100 million yen, the nationwide programs will cost about ¥70,000 billion yen. These numbers are significantly higher than the government capacity.

The above figures show the financial challenges in infrastructure maintenance. Investments on infrastructure maintenance are still inadequate, especially among local governments. Fig. 12-8(b) shows the proportion of expenditure by local government for infrastructure maintenance—in this case for bridges—as the ratio to the total assets. One can see that many local governments spend less than one percent, or even less than 0.5 percent of the total assets. The decrease in maintenance expenditure is expected to remain in the following years due to the tight budget. This shows a grim picture of our infrastructure systems that are not yet well maintained, and therefore the risks of infrastructure failure are increasing steadily in the future.

12.5 Monitoring of Infrastructures for Risk Reduction

Experiences from past disastrous events involving infrastructures have shown that prior to the accidents some damage indicators emerged. The post-failure investigation of the sudden collapse of the pedestrian bridge at the Paris-Charles-de-Gaulle airport in 2004, for instance, reveals that prior to the accident some unusual noise was created by the structure and dust was coming out of the concrete. Had these indicators been well monitored, the total collapse of structures might have been avoided. Capturing damage indicators—or warning signs, however, is not an easy task. Sometimes the warning signs are masked by other features, so that they become clear only if we closely monitor the structures. This fact emphasizes the need for preparedness through a continuous monitoring. The "Thatcher Law" that says: "*The unexpected happens, and you'd better prepare (be ready) for it*" might well be applied here.

12.5.1 Monitoring as a Means of Vulnerability Quantification

In the previous section, we have defined a risk in general as the product of probability of occurrence or hazard multiplied by the consequence. A similar definition is employed to structural analysis, where the risk is defined as the function of two quantities: hazard and structure vulnerability, in the form of:

$$Risk = f(\text{Hazard} \times \text{Vulnerability}) \qquad (12\text{-}6)$$

In this case hazard refers to the likelihood of occurrence of unexpected events that may endanger the structure. Structural vulnerability refers to the susceptibility of a building when subjected to a certain type of hazard. It depends on the structural configuration such as the system redundancy and ductility, as well as the materials and the quality of construction. The vulnerability will increase when the structure is aging, deteriorating, or when improper design/construction and initial defects appear. Equation (12-6) implies that the risk of structural failure depends on both quantity of hazard and vulnerability of structure. When the hazard is large but the structure is safe then the risk is small. On the other hand, if the structure is vulnerable, even a small hazard can increase the risk potential significantly. This fact emphasizes the need to identify and quantify structural vulnerability as accurately as possible.

In the previous section we defined the use of the fragility curve as an approach to quantify structural vulnerability. The empirical fragility curve, on one hand is very useful since it consists of real damage data due to specific events. On the other hand, however, to construct a statistically reliable curve one needs to collect large amount of data. Due to limitation of damage and natural hazard events, an analytical-based vulnerability estimation, such as the analytical fragility curve, has been developed. This approach has some significant limitations. For instance, the fragility curve is based on a normal distribution assumption, but it is often unclear where the exact position of the studied structure is in the assumed distribution.

Also, when conducting the statistical vulnerability analysis many assumptions regarding the structural conditions are employed: for instance, boundary condition, material properties, and locality effects. These assumptions usually rely on ideal conditions, while in fact the condition may be different. Furthermore, one needs to realize that civil structures, even though they appear typical, are not completely identical. They differ in boundary conditions, material qualities, soil condition and environment. These differences create uncertainties, and therefore some random effects must be taken into consideration. To minimize the uncertainties, a comprehensive analysis using numerical simulation or even experiment is needed. The two approaches, however, have equal difficulties. The numerical simulation approach requires information of the actual structural properties, while the experiment approach is even more difficult considering the inertia and the large scale of the structure. To this end, field monitoring is the only feasible option. Using monitoring the actual structure condition can be assessed and therefore its vulnerability can be quantified.

12.5.2 Sensor System and Monitoring Network: Current Practices and Challenges

Monitoring of hazard and structure vulnerability requires hardware and software. In dealing with monitoring of civil infrastructure, both hardware and software present challenges and difficulties. The problems regarding hardware are the selection of proper instrumentation or devices, system arrangement, and system representation. Two main factors contribute to the selection:

1. Scalability. As mentioned in the previous section, in term of scale, civil infrastructures lie within two extreme boundaries: the macro-scale extreme and the microscale extreme. This also applies to the selection of hardware for monitoring. At one end, when it comes to a macroscale such as the global scale, sensors linked to satellites in the global positioning system (GPS) are used. On the other hand, for the microscale problem, sophisticated nano or micro sensors are more suitable. The civil infrastructure scale is considered on the mes-oscale level. Because of the spatial diversity and sheer numbers of structures, the use of a too-expensive and sophisticated microsensing system should be avoided. At the same time, however, the uniqueness of information of each structure system should be retained.

2. Durability. Civil infrastructures are designed to last for a long time, typically over 50 years. During the service life, hazardous events may not occur very frequently, but nevertheless the moni-toring system must be always up and running. This indicates the need for not only reliable but also durable sensors.

In addition to hardware problems, software problems such as diagnostic methodologies and reliability of vulnerability assessment and performance evaluation also need to be solved.

Despite all challenges to the sensor system and monitoring network, a number of hazard monitoring systems are now operating in Japan. Some of the high-profile systems are:

- Kyoshin-Net (K-NET). In the aftermath of the 1995 Kobe Earthquake a comprehensive seismic hazard monitoring system was developed. The system constitutes a network of more than 1,000 wideband seismometers installed nationwide (Kinoshita (1998)).
- AMeDAS. A high-resolution surface monitoring system involving wind, temperature and other meteorological data was developed by Japan Meteorological Agency (JMA). The Automated Meteorological Data Acquisition System (AMeDAS) system consists of about 1,300 stations with automatic observation equipment. These stations, of which more than 1,100 are unmanned, are located at an average interval of 17 km throughout Japan and already in place, having started from 1974 (JMA-AMeDAS).
- SUPREME. Recently, as part of a seismic disaster-prevention system, Tokyo Gas installed a real time monitoring system named

Super Dense Real-Time Monitoring of Earthquakes (SUPREME) that consists of 3,900 sensors. These networks of sensors would stop gas distribution when a certain level of earthquake occurs. Installation of the upgraded sensor network was completed in March 2006 and the system is now effectively in place (Shimizu et al. (2006)).

- TERRA-S. Japan bullet train (Shinkansen) also has a seismic disaster prevention system. The Tokaido Shinkansen Earthquake Rapid Alarm System (TERRA-S) detects the P-wave at the immediate K-net stations and determines the impact of the seismic wave to Shinkansen and railway infrastructure. When the wave is considered hazardous, an alarm is sent to the running train using a much faster wave, and train operations are stopped prior to the arrival of the actual seismic wave. Investment for upgrading the system costs around US$6.3 million (JR Central (2005)).

The SUPREME and TERRA-S systems, in particular, are excellent examples of practical application of the real-time hazard monitoring systems. In application of TERRA-S system for instance, once the train stops, visual inspection is conducted to ensure that the railway structure is safe. This inspection takes few or several hours and depends heavily on human resources. Later, impact tests using weights of 30 kgf is performed to identify the natural frequencies of the viaduct. After performing inspections, train operation resumes when the structure's condition is considered safe.

The following are examples of the structure vulnerability monitoring systems that are now operating in Japan, including the ones that proposed by the authors:

- Doctor Yellow. For continuous monitoring of the railway system, Shinkansen employs a special inspection train nicknamed Doctor Yellow. The train is equipped with accelerometers and operates every ten days. By observing acceleration records, track irregularities and any structural defects as indicated by a large spike on the response, engineers can take subsequent countermeasures.
- TIMS. Having limited operational budgets, local railways operators, however, cannot implement the sophisticated and high-tech maintenance strategies in their facilities. Therefore, they require an accurate, simple, but low-cost maintenance system. To fill the need, the Train Intelligent Monitoring System (TIMS) was developed (Ishii et al. (2006)). The system consists of: (1) a triaxal accelerometer installed on the wagon floor to measure train acceleration,

(2) a Global Positioning System (GPS) sensor to locate train position and, (3) a portable computer to record position and response data. The measurement system is simple and reliable for detecting any irregularities on the surface of railway track. Such a portable, compact, and uncomplicated system makes it suitable for application on ordinary railway wagons at low-cost and with flexible time.

- VIMS. A monitoring system similar to the TIMS was introduced to monitor highway pavements and expansion joints. The Vehicle Intelligent Monitoring System (VIMS) (Fujino (2005)) utilizes dynamic response of a VIMS-installed car to capture the condition of road-pavement surfaces as well as the expansion joints. The system is proposed as alternative to the conventional road profiler system, whose operational cost is expensive. The VIMS system consists of: (1) an accelerometer to measure the dynamic response of the car, (2) a GPS to identify the position where the dynamic response is recorded, and (3) a portable computer to store the measurement data. Using the acceleration response and the correlation information, one can identify the conditions of the expansion joints. Furthermore, by comparing the amplitude of dynamic response and its correlation, one can continuously monitor the condition as well as any possible defects on expansion joints.

The examples on hazard and vulnerability monitoring systems listed above show that while hazard-monitoring systems are now very advanced, the vulnerability-monitoring systems are lagging behind. There are still many structural vulnerability-monitoring practices that depend heavily on human resources.

12.6 Conclusions

The paper has presented the state of the art of risk assessment and management of infrastructure systems. In assessing the risk, it is important to identify and quantify the hazard, vulnerability, and consequences. Examples of methodologies and current practices to identify and quantify hazards, vulnerability,and consequences of infrastructures have been presented. The paper also discusses the importance of risk assessment in the framework of risk management. The concept of risk assessment is implemented to evaluate the seismic, wind, and snow risk of civil infrastructure. Risk assessment is divided into two main parts: the hazard assessment and the structural-vulnerability assessment. Based on risk assessment, the risk comparison study is conducted to facilitate rational decision making.

The second part of the paper discusses the characteristics of urban infrastructures and the challenges they face. It is shown that the accumulations of urban infrastructures have created several difficulties regarding their management and sustainability. Several issues regarding infrastructures maintenance were discussed with examples. Regarding natural disasters, the paper has shown the rationality of investing in disaster prevention activities in reducing the number of casualties. It is realized that investments on infrastructure maintenance are still inadequate. Hence, the sustainability of infrastructure in the future is at risk unless we start increasing the investment in maintenance from now.

Based on the lessons from previous natural disasters, Japan has now developed comprehensive hazard-monitoring systems nationwide. The systems have proved to be beneficial in assessing and communicating the risk and also facilitating decision making. A complete risk-assessment system should incorporate the hazard and vulnerability assessments. The latter aspect, however, still lags behind and requires further research efforts. The concept of structural-health monitoring as a tool for vulnerability assessment is therefore proposed.

References

Bernstein, P.L. (1996) *Against the Gods–The Remarkable Story of Risk.* New York: Wiley

Fujino, Y. (2005) *Development of Vehicle Intelligent Monitoring System (VIMS)* (SPIE International Symposia on Smart Structures & Materials/NDE, Conference #5765)

Hayashi, M. (2000) *For Scientific Study on Risk Concept.* Kougakuin University Press, Tokyo, Japan 38(1)

Hirota, S. (2002) *Risk World–Psychological Perspective.* Tokyo: Keio University Press

Ishii, Y., Fujino, Y., Mizuno, Y. and Kaito, K. (2006) *The Study of Train Intelligent Monitoring System Using Acceleration of Ordinary Trains,* in Proceedings of Asia-Pacific Workshop on Structural Health Monitoring, Yokohama Japan (CD)

ISO/IEC (2002) Guide 73 risk management—vocabulary—guidance for use in standard

Japan Meteorology Agency-AMeDAS (http://www.jma.go.jp/jp/amedas/) (Accessed on 12 February 2008)

J.R. Central (2005) *Operation Launch of the New Earthquake Rapid Alarm System "TERRA-S" for Tokaido Shinkansen,* http://jr-central.co.jp/eng.nsf/english/n-05-0905 (Accessed on 12 February 2008)

Kinoshita, S. (1998) "Kyoshin net (K-net)", *Seismological Research Letters*, 69(4): 309–332.

Kondo, K., Kanda, J. and Choi, H. (2002) "Study on Strong Wind Hazard Analysis for Buildings", *Proceedings of National Symposium on Wind Engineering*, 17:191–196

Lewis, H.W. (1990) *Technological Risk*. New York: W.W. Norton

Murao, O. and Yamazaki, F. (2000) "Fragility Functions of Houses Based on Municipal Investigation Date of the 1995 Kobe Earthquake", *Journal of Architectural Institute of Japan (AIJ)*, 527:189–196

National Research Institute for Earth Science and Disaster Prevention (1982) *Report on Damage of Hokuriku Region Due to the 56 Heavy Snowfalls. Report of Major Disaster No. 17. Tokyo*. Tokyo: NIED

Shimizu, Y. et al. (2006) "Development of Real-Time Safety Control System for Urban Gas Supply Network", Journal of Geotechnical and Geoenvironmental Engineering 132:237–249

Stewart, M.G. and Melchers, R.E. (1997) *Probabilistic Risk Assessment of Engineering Systems*. London: Chapman & Hall

Takada, T. (2006) *Risk Comparison of Natural Hazards in Japan*. Glasgow: ASRANET-CDROM

United States Nuclear Regulatory Commission (1981) *A Risk Comparison*. Washington, DC: NUREG/CR-1916

13. Resource Recycling in Concrete: Present and Future

Takafumi Noguchi

13.1 Resource Input and Waste Output in Construction

According to a White Paper on the Environment (Ministry of the Environment (2003)), the total material input of Japan ranged from 2.0 to 2.2 billion tons annually in recent years, of which 1.0 to 1.1 billion tons (50 percent) were accumulated every year in the form of buildings and civil structures as shown in Fig. 13-1(a), which indicates the enormous consumption of resources by the construction industry compared with other industries. The production of concrete, a primary construction material for forming the infrastructure of modern nations, amounted to approximately 500 million tons (217.4 million cubic meters volume, converted by assuming the density to be 2.3) in 2000 in Japan, accounting for nearly 50 percent of the annual resource consumption of the construction industry as shown in Fig. 13-1(b). In other words, concrete accounts for nearly 25 percent of Japan's total material input. Incidentally, the construction industry's 2001 consumption of steel and wood, two other primary construction materials, amounted to 32,530,000 and 17,000 t (4,170,000 and 34,000 m³ volume, converted by assuming the density to be 7.8 and 0.5), respectively, both far less than concrete consumption.

Furthermore, the amount of waste in Japan totaled approximately 458,360,000 t in 2000 (general waste: 52,360,000 t; industrial waste: 406,000,000 t) as shown in Fig. 13-2(a) (Ministry of Land, Infrastructure and Transport (2002)). Waste from construction accounted for approximately 20 percent (79,000,000 t) of total industrial waste. Moreover, in 2000, construction waste accounted for nearly 30 percent (12,800,000 t) of the 45,000,000 t of industrial waste destined for final disposal sites and approximately 60 percent (241,000 t) of the 400,000 t of illegally dumped industrial waste. As concrete lumps account for approximately 42 percent (35,000,000 t) of total construction waste as shown

Y. Fujino, T. Noguchi (eds.) *Stock Management for Sustainable Urban Regeneration*,
© 2009 to the complete printed work by Springer, except as noted. Individual authors
or their assignees retain rights to their respective contributions; reproduced by permission.

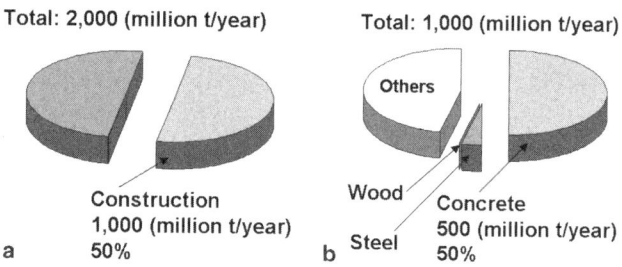

Total: 2,000 (million t/year)   Total: 1,000 (million t/year)

Construction   Wood   Concrete
1,000 (million t/year)   500 (million t/year)
a   50%   b   Steel   50%

Fig. 13-1. (a, b) Resource input into construction industries

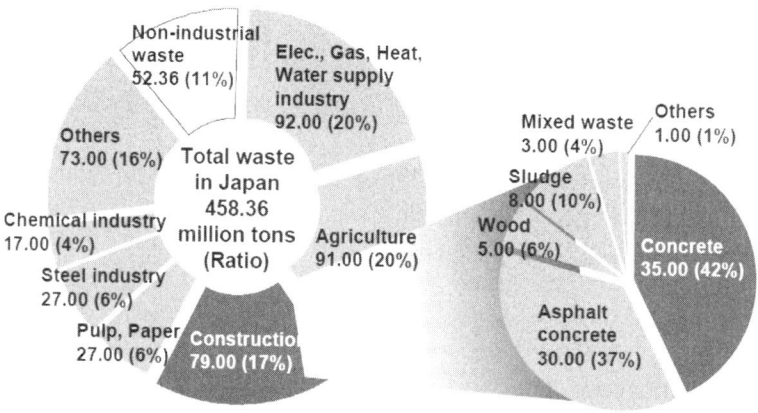

Fig. 13-2. (a, b) Waste output from construction industries. **a** total waste. **b** construction waste

in Fig. 13-2(b), approximately 8 percent of total waste in Japan therefore consists of concrete lumps.

As stated above, concrete accounts for large percentages of both resource input and waste discharge. Thus promotion of the recycling of demolished concrete is a pressing social issue in Japan where the remaining capacity of landfill sites for industrial waste is diminishing every year as shown in Fig. 13-3 (Ministry of the Environment (2003)). The shortage of landfill space in Japan, especially in urban areas, is really serious.

13.2 Industrial Waste into Concrete

Industrial waste and byproducts including slag and coal ash have long been utilized as cement materials, as these contain significant amounts of major ingredients of cement. Combustible waste including waste tires

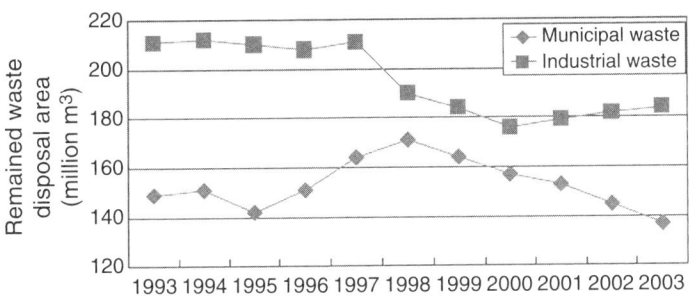

Fig. 13-3. Waste disposal area

and waste plastics has also been used as fuels for cement production. Recently, sewage sludge and incineration ash from municipal solid waste have been actively utilized as raw materials and fuel for cement because (1) the high temperature safely decomposes hazardous materials including dioxins; (2) no new waste is generated; (3) this requires no additional incineration facilities; and (4) it extends the service lives of final disposal sites. Fig. 13-4 and Fig. 13-5 show the waste and byproducts utilized by the cement industry and the state of utilization of blast-furnace slag and coal ash by the cement and concrete area in Japan (Japan Cement Association (2008)), respectively.

"Ecocement" made using incineration ash from municipal solid waste and sewage sludge as its main and secondary materials, respectively, was recently developed and the related Japan Industrial Standard, JIS R 5214 (Ecocement) was established in 2002. The raw materials of Ecocement and chemical compositions of incineration ash are shown in Fig. 13-6. Its production processes are basically the same as those for normal cement but are characterized as follows:

1. Since dioxins completely decomposed at a high temperature are rapidly cooled to below 200°C in a cooling tower, they are captured in a bag filter and an active coke-packed tower without being resynthesized.
2. Trace heavy metals contained in incinerated ash are recovered and concentrated, and delivered to nonferrous metal factories for separation of copper, lead, zinc, etc., for recycling.

Aggregate, which occupies 70 percent of concrete by volume, is anticipated as a recipient of waste and byproducts from other industries, with various types of recycled aggregates being developed as given in Table 13-1. However, such recycled aggregate still accounts for only a small fraction of the total aggregate production for concrete.

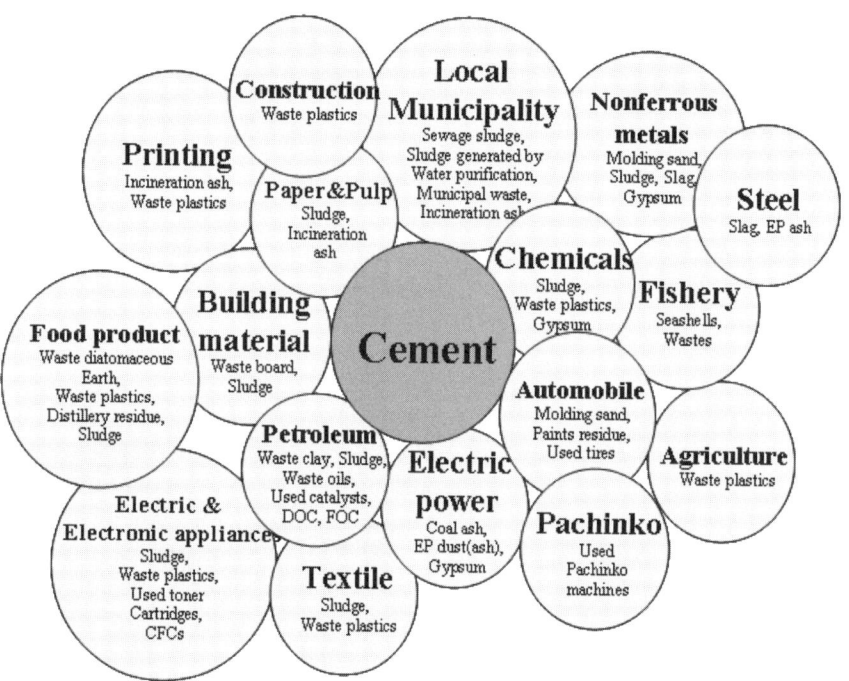

Fig. 13-4. Utilization of byproducts from other industries for cement production

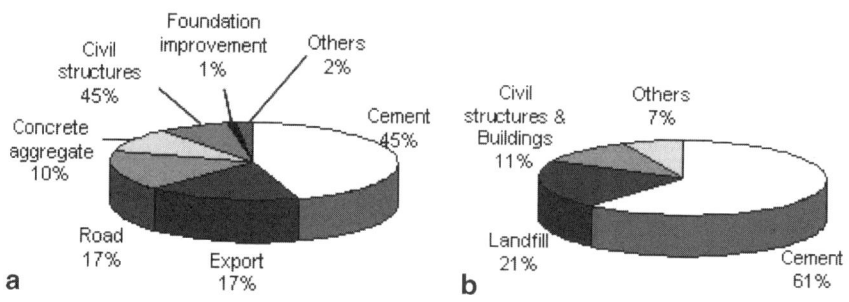

Fig. 13-5. (a, b) Utilization of blast furnace slag and fly ash for cement and concrete. A blast furnace slag (24,518,000 t). B fly ash (6,919,000 t)

The concrete industry, which is a powerful waste management route because of its capacity and the possibility of using various wastes from other industries, can be located at the core of the recycling ecosystem. However, the maintenance of safety and quality of products must be established as a principle of the concrete industry. Further voluntary agreements are required, and societal understanding will support the growth of recycling activities.

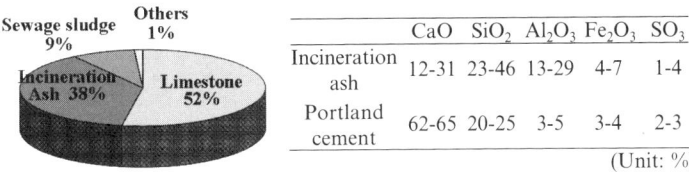

Pie chart: Sewage sludge 9%, Others 1%, Incineration Ash 38%, Limestone 52%

	CaO	SiO$_2$	Al$_2$O$_3$	Fe$_2$O$_3$	SO$_3$
Incineration ash	12-31	23-46	13-29	4-7	1-4
Portland cement	62-65	20-25	3-5	3-4	2-3

(Unit: %)

Fig. 13-6. Raw materials of Ecocememt and chemical compositions of incineration ash

Table 13-1. Various types of aggregates made from industrial waste

Type	Standard
Blast furnace slag aggregate	JIS A 5011–1
Ferronickel slag aggregate	JIS A 5011–2
Copper slag aggregate	JIS A 5011–3
Electric arc furnace oxidizing slag aggregate	JIS A 5011–4
Melt-solidified aggregate for concrete derived from municipal solid waste and sewage sludge (Molten slag aggregate for concrete)	JIS A 5031
Fly ash calcined lightweight aggregate	

13.3 Concrete Recycling

13.3.1 Destinations of Demolished Concrete

With the aim of solving the construction waste problem, the Japanese Ministry of Land, Infrastructure and Transport (MLIT, formerly the Ministry of Construction) formulated an Action Plan for Construction Byproducts (Recycling Plan 21) in 1994, which called for halving the amount of final disposal of construction waste by 2000, and a Promotion Plan for Construction Waste Recycling in 1997, which includes principles, objectives, and measures for further promoting recycling of construction waste. Thanks to such active and continual policies, construction waste discharge began to decrease, with the recycling ratio of concrete lumps and asphalt concrete lumps exceeding 95 percent as shown in Fig. 13-7 (Ministry of Land, Infrastructure and Transport (2006)). In view of the still low recycling ratios of waste wood, slime, and mixed waste generated by construction, the MLIT then enforced the Basic Law for Establishing a Recycling-based Society, the Construction Material Recycling Act, and the Law on Promoting Green Purchasing.

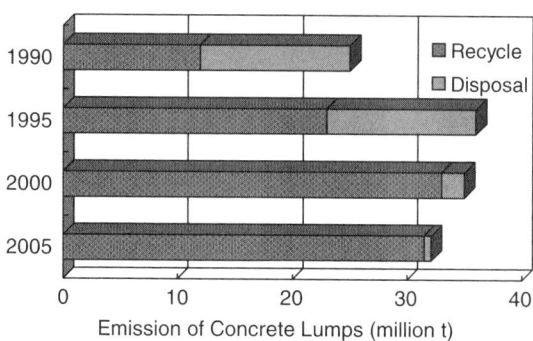

Fig. 13-7. Change in recycling ratio of concrete

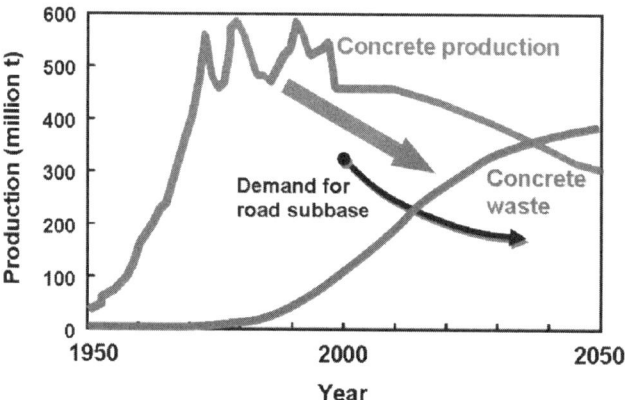

Fig. 13-8. Predicted concrete lumps

However, concrete lumps, which boast a high recycling ratio, are entirely destined for bottoming and grading adjusters for arterial high-standard highways, urban expressways, and general roads designated by the Road Bureau of the MLIT. The quality of recycling is therefore completely different from that of asphalt concrete lumps, for which level-cycling is accomplished. As shown in Fig. 13-8, an enormous amount of demolished concrete lumps will be generated in the near future from concrete structures mass-constructed during Japan's rapid economic growth era—structures that are doomed to demolition due to durability problems. Moreover, road construction is decreasing and the method of repair is expected to shift from replacing to milling and applying an overlay. These trends will lead to an imbalance between the supply of demolished concrete and the demand for road bottoming. Also, the volumetric reduction of future infrastructures based on population estimation and the extension of the service life of the existing stock by increased succession, utilization, and conversion will keep on reducing the amount of new construction of structures and concrete production. Accordingly, this will culminate in the need to recycle aggregate into aggregate for concrete, and it is no exaggeration to say that these recycled materials can account for the greatest part of future aggregate for concrete. It is therefore vital to convert recycling from quantity-oriented to quality-oriented recycling as proposed in the Promotion Plan for Construction Waste Recycling 2002 formulated by the MLIT in that year. In other words, it is necessary to find optimum recycling methods with due consideration to the material balance, while promoting the production and supply of high-quality recycled aggregate. It should also be noted that a large discrepancy exists between the amount of demolished concrete generated at demolition sites and the amount discharged from such sites. This may result from the reuse of demolished concrete as a material for road bottoming or backfill on site and its disposal as part of mixed construction waste. This discrepancy poses no problem if all the dissociation results from reuse on site, but if it includes mixed waste and illegal dumping, then the amount of discharged concrete may rise by a corresponding amount when such practices are rectified in the future. Recycling of demolished concrete as aggregate for concrete will then become even more compelling.

13.3.2 Technologies for Concrete Recycling

The uses for concrete lumps to be recycled are determined by the qualities of the recycled material, such as density and water absorption, which vary depending on the percentage of cement paste contained within or adhering to the surfaces of the original aggregate, and the quality of recycled aggregate depending on the production method. Fig. 13-9 shows general methods for producing recycled road bottoming, recycled aggregate for leveling concrete (low-quality recycled aggregate), and recycled aggregate having qualities comparable to those of natural aggregate and used for structural concrete (high-quality recycled aggregate). Single toggle-type jaw crushers are generally used for the primary crushing of demolished concrete regardless of the ultimate quality of recycled aggregate. Impact crushers are used for secondary and tertiary crushing when producing middle- and low-quality recycled aggregate. While the quality of recycled aggregate produced by using such equipment is improved as the number of treatment processes increases, the recovery percentage of recycled aggregate decreases with increased amounts of powder byproducts as the

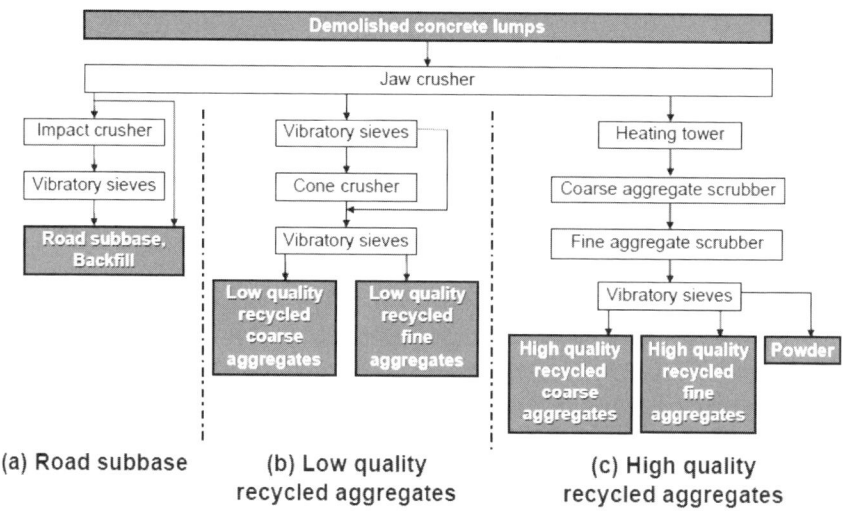

(a) Road subbase   (b) Low quality   (c) High quality
                          recycled aggregates   recycled aggregates

Fig. 13-9. Recycling process of concrete lumps

aggregate itself is crushed. Special equipment is therefore necessary for efficient production of high-quality recycled aggregate. Other equipment in practical use for producing middle- and low-quality recycled aggregate includes self-propelled or vehicle-mounted jaw crushers and impact crushers that save the energy normally expended to haul the demolished concrete.

Equipment in the stage of practical use for the efficient production of high-quality recycled aggregate is shown in Figs. 13-10(a–d). A heating grinder, shown in Fig. 13-10(a), incorporates a mechanism using a tube mill to separate aggregate from cement paste embrittled by heating concrete lumps roughly crushed to 40–50 mm in diameter to 300 °C. This is the only device capable of also producing recycled fine aggregate for concrete.

An eccentric rotor-type mechanical grinder, shown in Fig. 13-10(b), has a mechanism whereby concrete lumps charged into a space between the inner cylinder that rotates eccentrically and the outer cylinder are made to scrub one another to remove cement paste adhering to aggregate surfaces. It is capable of producing 60 t of recycled aggregate per hour. A screw mill, shown in Fig. 13-10(c), has a mechanism whereby concrete lumps charged into a cylinder having a twin cone are scrubbed by one another. This process is automatically repeated several times according to the required quality. A wet scrubber/levigator, shown in Fig. 13-10(d), incorporates a mechanism whereby concrete lumps crushed in multiple stages using a crusher and wet scrubber are moved up and down in water to separate mortar and wood chips with a low density from coarse aggregate.

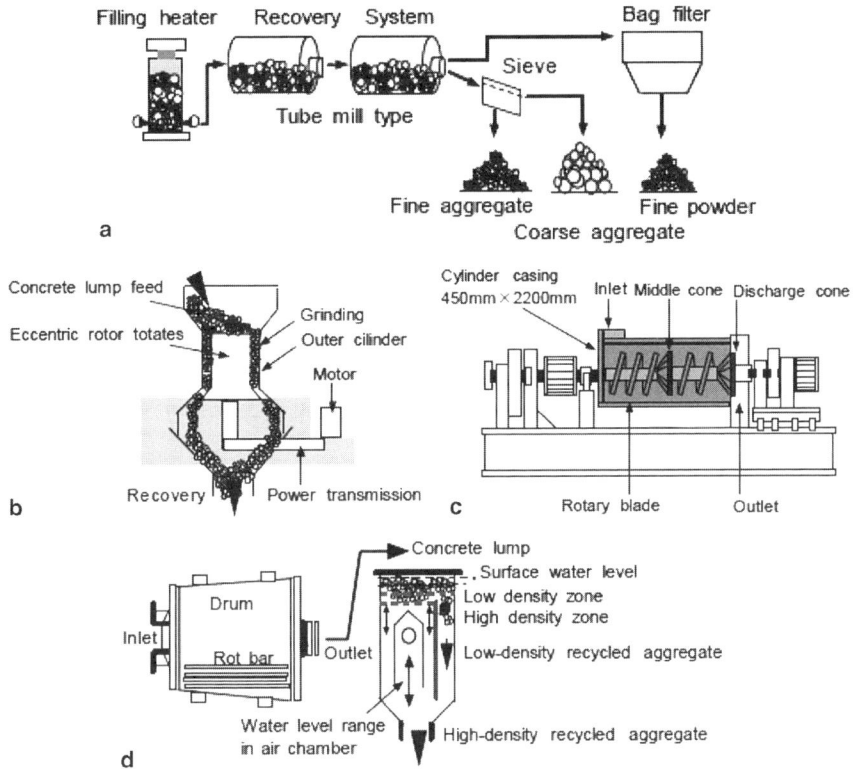

Fig. 13-10. (a–d) Equipment for production of high-quality recycled aggregate.
(a) Heated scrubbing method. (b) Mechanical scrubbing method: eccentric tubular
type. (c) Mechanical scrubbing method: screw type. (d) Wet scrubbing and gravity
classification method

Problems are posed by a large amount of by-product powders that result
from the production of high-quality recycled aggregate. Possible uses for such
powders considered so far include the following: addition to road bottoming,
a cement material, concrete addition, asphalt filler, ground improving material, and inorganic board material. However, these are in competition with
inexpensive natural resources, demanding quality stabilization and quality-
control cost reduction of such fines.

13.3.3 Standards for Recycled Aggregate for Concrete

Research and development aimed at using demolished concrete as recycled
aggregate for concrete began in the 1970s. Several draft standards have been

established, and the Standardization Committee for Recycled Aggregate was set up in the Japan Concrete Institute in 2002, which was tasked with formulating Japan Industrial Standards (hereinafter JIS) for recycled aggregate for concrete. The Committee established three JISs as follows:

- JIS A 5021 (Recycled aggregate for concrete, Class H, hereinafter RA-H)
- JIS A 5022 (Recycled concrete using recycled aggregate, Class M, with Annex [Recycled aggregate for concrete, Class M, hereinafter RA-M])
- JIS A 5023 (Recycled concrete using recycled aggregate, Class L, with Annex [Recycled aggregate for concrete, Class L, hereinafter RA-L])

Three types of recycled aggregate are classified by water absorption and oven-dry density, each being recommended for concrete structures and segments as given in Table 13-2. This classification urges a shift to a design system that permits the use of each class for suitable structures and segments. High-quality recycled aggregate is suitable for structures and segments requiring high durability and strength, while middle- to low-quality

Table 13-2. Physical properties requirements for RA

	RA-H		RA-M		RA-L	
	Coarse	**Fine**	**Coarse**	**Fine**	**Coarse**	**Fine**
Oven-dry den-sity (g/cm^3)	Not less than 2.5	Not less than 2.5	Not less than 2.3	Not less than 2.2	–	–
Water Absorption (%)	Not more than 3.0	Not more than 3.5	Not more than 5.0	Not more than 7.0	Not more than 7.0	Not more than 13.0
Material passing 75 μm sieve (%)	Not more than 1.0	Not more than 7.0	Not more than 1.5	Not more than 7.0	Not more than 2.0	Not more than 10.0
Scope of application	No limitations are put on the type and segment for concrete and structures with a nominal strength of 45 MPa or less		Members not subjected to drying or freezing-and-thawing action, such as piles, underground beam, and concrete filled in steel tubes		Backfill concrete, blinding concrete, and leveling concrete	

recycled aggregate, which can be produced with minimal cost and energy or powdery by-products, is suitable for other structures and segments.

The JIS A 5021 includes the following policies and requirements:

- RA-H is the high-quality aggregate used for JIS A 5308 (Ready-mixed concrete) in a similar way to natural good-quality aggregate.
- Demolished concrete lumps usually contain considerable amount of brick, tile, plaster, plastics, etc. The amount of deleterious substances in RA-H is tested through comparison of the RA-H thus produced with several confirmatory samples that include each deleterious substance. The limits of deleterious substances are shown in Table 13-3.
- The minimum rate of sampling and testing and the criteria of alkali-silica reactivity of RA-H are dependent on whether the attribute of original aggregates is identified. Unless it is identified, the produced RA-H must be sampled and tested at frequent intervals and treated as an aggregate that has potential alkali-silica reactivity.
- The RA-H must be produced in a plant where the quality of produced RA-H and the producer's production control are assessed and surveyed by an approved inspection body and then certified by an approved certification body.

A quality control system for construction materials has been established, to ensure that materials, particularly JIS products and products conforming to JIS, are supplied with constant qualities that meet the specifications under strict quality control, so that contractors and citizens can carry out construction and use the resulting structures with peace of mind. While it is essential to establish such a system for promoting the wide use of recycled aggregate, a distinctive difference exists between crushed stone/sand for concrete and recycled aggregate with regard to material procurement. Whereas crushed

Table 13-3. Limits of amount of deleterious substances in RA-H and RA-M

Category	Deleterious substances	Limits (mass%)
A	Tile, brick, ceramics, asphalt concrete	2.0
B	Glass	0.5
C	Plaster	0.1
D	Inorganic substances other than plaster	0.5
E	Plastics	0.5
F	Wood, paper, asphalt	0.1
Total		3.0

stone/sand that is deemed uniform to a certain extent can be procured in large quantities, the grading, density, water absorption, alkali-silica reactivity, etc., of demolished concrete may naturally vary from one lump to another, particularly when the recycled aggregate production plant is located away from demolition sites (offsite plant) and accepts demolished concrete from various structures. To promote recycled aggregate as JIS products or JIS-conforming products, it is therefore necessary to (1) produce recycled aggregate from only specific structures at onsite plants; (2) carry out material control by separately storing concrete lumps from each structure at offsite plants; or (3) carry out quality control by substantially increasing the frequency of acceptance inspections and product inspections at offsite plants.

13.3.4 Complete Recycling of Concrete

A conventional manufacturing system can be regarded as a system focusing on cost saving and efficiency without consideration of ease of disassembly of the products and component materials, based on which most existing concrete structures have been produced. Since the technology for recycling the products of such a system is inevitably required to treat used products not intended for recycling at the design stage, recycling can be regarded as an alternative for waste disposal. This is a typical end-of-pipe approach leading to downcycling whereby materials are diffused into a wide range of industries. Recycled concrete produced in this manner are seeded with potential problems given in Table 13-4 (Tamura et al. (2002)). Fig. 13-11 shows the material flow of concrete structures produced by the existing forward process systems (Tamura et al. (2002)). Unless the high-quality recycled aggregate production system is applied, no closed loop is formed to circulate resources using only concrete structures. Natural resources constituting the structures turn into nonstructural concrete materials and then into recycled crushed stone for road subbases in the course of quality degradation through downcycling as part of an open-loop form of circulation. Since natural resources are also input to byproducts of this cycle, final disposal-type output is produced in abundance, thereby aggravating environmental disruption.

But this is still essential, as the enormous stock of existing concrete structures will somehow demand treatment of concrete lumps in the future. If recycled products happen to be in great demand with a low level of performance requirements, downcycling can be an effective solution for a certain period until such demand disappears. Road bottoming can be regarded positively as an instance of such a solution. Despite continuous research and development

Table 13-4. Problems of regenerated concrete based on forward-process production systems

Issues	Contents
Quality: degrade	The performance of recycled concrete generally degrades from the original concrete because the performance of recycled aggregate naturally deteriorates when no consideration is given to inverse processes at the stage of material design. The resulting recycling system reduces the waste output but cannot restore the performance/value of the regenerated concrete.
Supply: unstable	The regenerated concrete does not use recycled aggregate stockpile for maintenance but uses recycled aggregate generated from existing structures. The recovery inevitably fluctuates, and the supply of regenerated concrete becomes unstable.
Price: unstable	The market price of regenerated concrete is determined by comparison with the original concrete. It is therefore strongly affected by the cost fluctuation of the original concrete regardless of the processing cost. Due to the above-mentioned quality degradation, the price tends to be reduced.
Distribution: unstable	Solution to fundamental problems is essential for realizing a continuous balance between supply and demand and steady circulation/production of regenerated concrete having unstable quality, supply, and price. At present, It is extremely difficult to establish a closed circulation system of regenerated concrete.
Environmental impact: increase	Setting up an inverse process system and carrying out rational operation of recycled products while maintaining existing forward process production systems will potentially have a strong impact on the environment. A view will then be voiced frequently that disposal is more appropriate than recycling in terms of energy consumption in consideration of sustainability of the global environment.

since the early 1970s, structural recycled aggregate has not been actively used. This is presumably due to the absence of application development based on social needs for the product or technological development in view of equitability between generations.

In consideration of the problems inherent in the conventional end-of-pipe approach (the open-loop system as shown in Fig. 13-12(a)), a new solution should be an integrated inverse manufacturing system, a closed-loop system

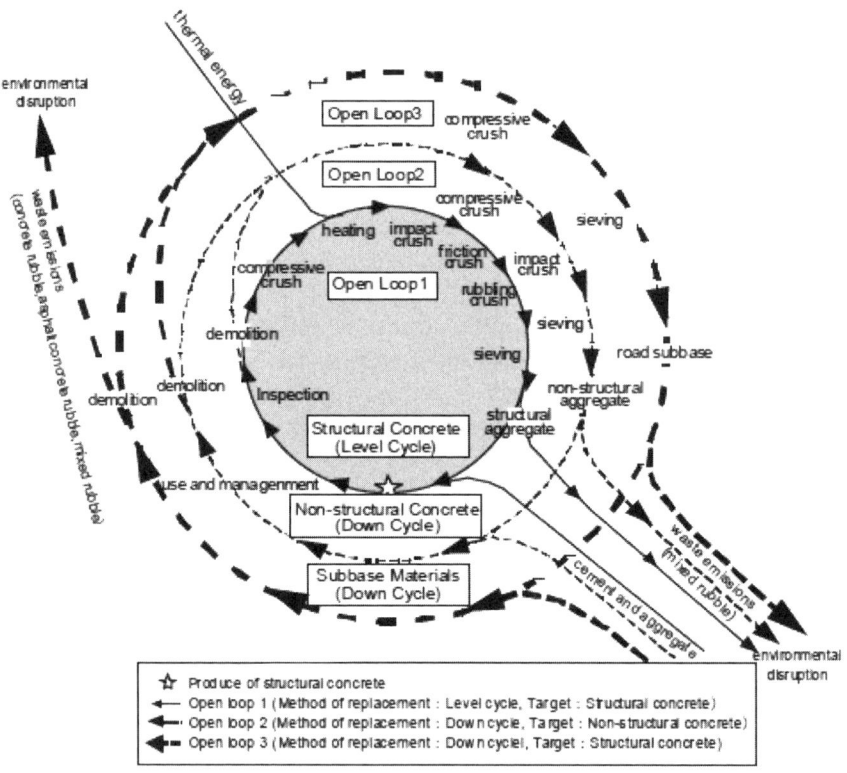

Fig. 13-11. Open-loop material flow of conventional concrete

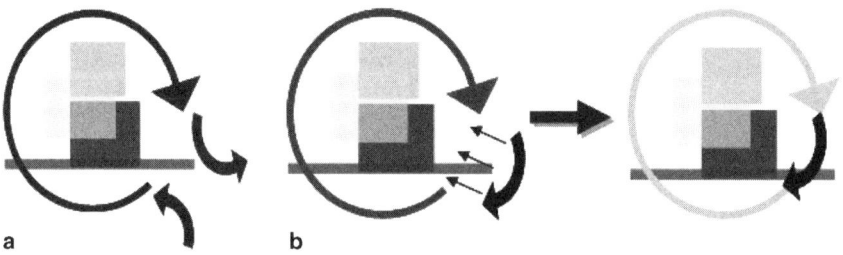

Fig. 13-12. (a, b) Manufacturing system. (**a**) Open-loop. (**b**) Closed-loop

defined as a manufacturing system in which downstream processes are consistent with upstream processes for the purpose of ensuring resource circulation using component materials that can be assembled and disassembled by similar work loads as shown in Fig. 13-12(b). This system is characterized by ease of disassembly as well as assembly incorporated at the design stage.

Specific methods of completely recycling the entire amount of concrete by applying recycling design incorporating inverse processes are described here. Maintenance design for durability retention of structural bodies and reuse design of members by skeleton-infill systems are excluded here. The principle of recycling design to realize complete recycling is, as stated above, to apply material design that proactively controls generation of waste and facilitates resource circulation in closed loops. When actually designing structures, it is important to ensure resource conservability of concrete materials: the property of materials to be able to continue circulation in different products as the media, with which they are used as components of concrete during its service life and, after demolition/separation, can be used as materials for products having the same or higher qualities, since the performance as components of the products are retained intact through demolition/separation.

Cement recovery-type completely recyclable concrete (Tomosawa and Noguchi (1995)) is defined as "concrete whose materials are entirely usable after hardening as materials of cement or recycled aggregate, since all the binders, additions, and aggregate are made of cement or materials for cement." It can formulate a closed-loop circulation material flow as shown in Fig. 13-13(a). Conversion from conventional concrete to completely recyclable concrete will substantially mitigate the environmental problem of concrete waste generation and CO_2 emission during cement production,

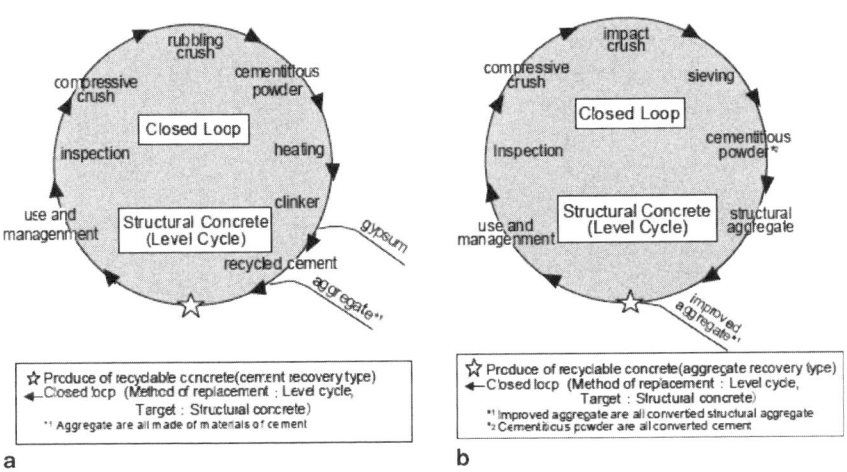

a   b

Fig. 13-13. (a, b) Closed-loop material flow of completely recyclable concrete. (**a**) Cement-recovery type. (**b**) Aggregate-recovery type

while permanently preserving and storing the limestone resource in the form
of structures. By adequately combining byproducts from industries other
than construction, such as blast-furnace slag and fly ash, it is also possible
to produce "completely recyclable concrete requiring no composition
adjustment," which can be used as a cement material as-crushed after
demolition without adding any other components. With the strong demand
for promotion of structures that reduce the environmental impact by reducing
waste, completely recyclable concrete is considered to be an essential
material for such structures.

Aggregate recovery-type completely recyclable concrete (Noguchi and
Tamura (2001)) is defined as "concrete in which the aggregate surfaces
are modified without excessively reducing the mechanical properties
of the concrete, in order to reduce the bond between aggregate and the
matrix, thereby permitting easy recovery of original aggregate." It can
form a closed circulating material flow as shown in Fig. 13-13(b). In
order to achieve 100 percent circulation of concrete in a closed system,
a structure is necessary as an aggregate-supplier in addition to one as
a cement-material supplier. By building a stock of structures keeping
such an appropriate balance, all cement and aggregate can be exploited
from built structures in the future. In aggregate recovery-type completely
recyclable concrete, aggregate surfaces can be modified by two methods,
i.e. chemical treatment and physical treatment. Chemical treatment is a
method in which formation of cement hydrates on the aggregate-paste
interfaces is chemically restricted, whereas physical treatment forms a
film to smooth the fine irregularities on the interfaces, thereby reducing
the mechanical friction. Modification treatment can be carried out simply
and economically by either method. Fig. 13-14 gives the aggregate surface
modification methods employed to easily recover original aggregate from
concrete (Tamura et al. (2002)).

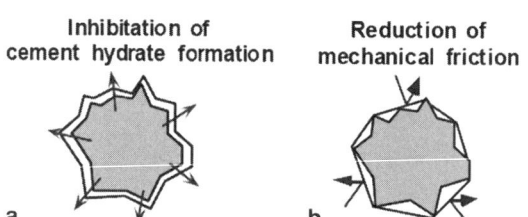

Fig. 13-14. (a, b) Improvement methods for surface of aggregate. (**a**) Chemical
treatment. (**b**) Physical treatment

13.4 Toward a Veritable Resource Recycling Society

13.4.1 Principles of Recycling

As stated above, conventional concrete recycling through low-quality recycled aggregate has yet to be marketable, whereas recycled aggregate having qualities equal to normal aggregate can be used for general structural concrete, though with drawbacks related to the cost.

Promotion of recycling is essential for establishing a recycling-oriented society. However, it is difficult for low-quality products to achieve marketability simply because they are recycled, as illustrated in Table 13-4.

- Recycling technology should fulfill the following principles:
- Recycling should be of high quality and
- Recycling should be repeatable

The former means that recycled products are not marketable unless they are of a quality that satisfies users, and recycling is not a viable proposition in spite of the accurate description as a "recycled product." Low quality of recycled products should be considered to indicate the immaturity of the recycling technology and the need for technology improvement or new technology development to achieve quality recycled products.

The latter means that, if a recycled product has to be dumped in a landfill after use with no chance of recycling, then the recycling is no better than producing waste of the following generation, contradicting the formulation of a recycling-oriented society. Such a product also burdens the purchaser with the responsibility of waste disposal. Potential purchasers will hesitate to purchase the product the moment they realize this pitfall, preventing the recycling loop from closing (Old Maid rule). Accordingly, to be repeatable, recycling must aim for the reproduction of the same product in the original sense of the word to form a loop.

Care should be exercised regarding so-called cascade recycling, mixed/compound recycling, recycling into other industries, and byproduct utilization, as these tend to be unrepeatable. When a product utilizing a byproduct is repeatedly recycled, the byproduct as a material will become useless, ending up as waste. What is a truly recycling-oriented society? Humans live on a cycle of resource use in which they take various resources from nature into their society, obtain benefits from them for a given period by processing and using them, and return them back to nature as waste when they cease to be useful. Once taken into human society, the resources are altered or modified and by the time they are returned to nature for final disposal, their state has changed to a certain extent. Just one or two recycling phases represent a short stay of resources in human society, which ends up with disposal (return) to nature. True recycling should eliminate such scenarios. Unrepeatable recycling merely extends their stay on the human side without contributing to the basis for a truly recycling-oriented society. The only acceptable case for this type of cycle is the use of resources in such a way that the product returned to nature does not represent an environmental load. A recycling-oriented society in the true sense of the word is a society that continues to use resources, once they are taken from nature into the society, without returning them unless they do not represent an environmental load. In such a society, the intake of resources from nature is minimized and products and materials that cannot be recycled repeatedly are rejected. While this vision may be overly idealistic, it should be kept in mind when evaluating and developing recycling technology.

13.4.2 Optimization of Material Flow

Fig. 13-15 shows the current material flow of concrete for one year in Japan. The discharge of resources related to concrete is as low as 10 percent of the input, leaving enormous amounts of concrete structures in stock. Because the figure does not demonstrate a material flow for a long period of time corresponding to a general service life of concrete structures, but only that for a given year, changes in industrial structure and business custom are not taken into account. The real material flow is, however, built upon a very complex society that

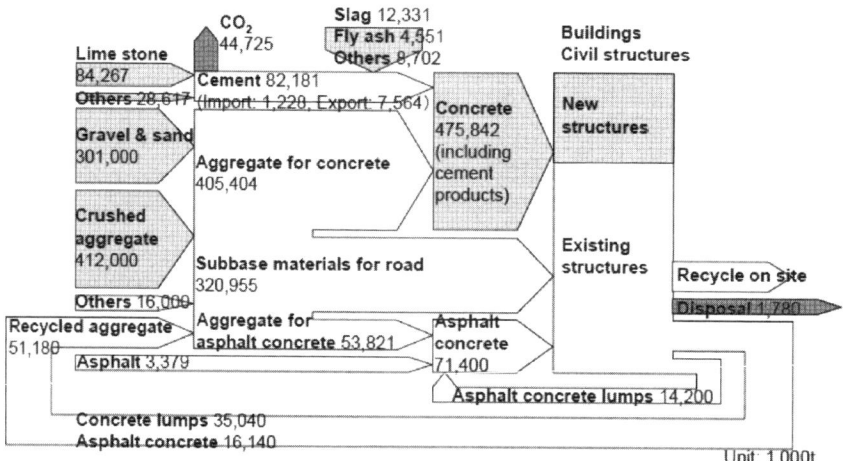

Fig. 13-15. Flow of concrete-related materials

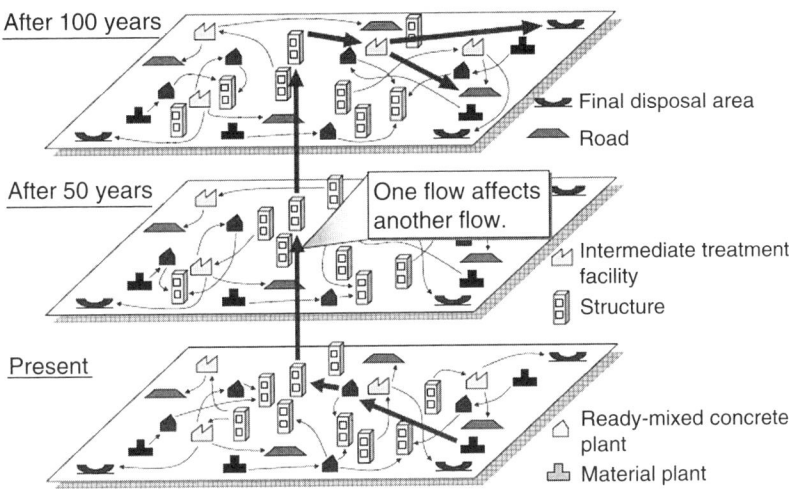

Fig. 13-16. Material flow in real society

contains geographic spread and temporal overlap as shown in Fig. 13-16. There also exists regionally-specific system and practices, and various companies with different principles in the real society. Therefore an optimized material flow for a given time or a given concrete structure does not always express an optimized material flow for society as a whole. For example, utilization of slag cement in consideration of reduction of CO_2 emission in the area where blast furnace slag is not produced needs surplus energy for transportation, resulting in an increase of CO_2 emission. While a utilization of concrete lumps for road bottoming materials may be a suitable application to restrain the consumption of natural resources and to reduce the waste, it may not lead to an optimum material flow for the next several decades to 100 years.

Consequently, an optimum flow has to be established in concrete-related materials, taking management principles and policy responses of individual enterprises into consideration based on actual social conditions and industrial structure in order to establish a real resource recycling society in which environmental impact is minimized as much as possible.

References

Japan Cement Association (2008) *Acceptance Situation of Waste and Byproducts*, http://www.jcassoc.or.jp/cement/1jpn/jg2a.html (16 March 2008) (in Japanese)
Ministry of the Environment (2003) *White Paper on Recycling Society*, http://www.env.go.jp/policy/hakusyo/junkan/h18/html/jh0601010400.html#3_2_4_1 (26 October 2007) (in Japanese)

Ministry of the Environment (2003) *Annual Report on the Environment in Japan 2003*, http://www.env.go.jp/en/wpaper/2003/index.html (12 November 2006)

Ministry of Land, Infrastructure and Transport (2002) *Report of Survey on Construction By-Products in Japan 2002*, http://www.mlit.go.jp/sogoseisaku/region/recycle/pdf/fukusanbutsu/jittaichousa/H14sensus.pdf (12 November 2006) (in Japanese)

Ministry of Land, Infrastructure and Transport (2006) *Report of Survey on Construction By-Products in Japan 2005*, http://www.mlit.go.jp/sogoseisaku/region/recycle/pdf/fukusanbutsu/jittaichousa/H17sensus.pdf (16 March 2008) (in Japanese)

Noguchi, T. and Tamura, M. (2001) "Concrete Design toward Complete Recycling", *Structural Concrete*, 2(3): 155–167

Tamura, M. et al. (2002) *Life Cycle Design Based on Complete Recycling of Concrete*, in Proceedings of the First Fib Congress, Osaka, Japan, October 2002, Concrete Structures in the 21st Century 2 (Session 8): federation internationale du beton.

Tomosawa, F. and Noguchi, T. (1995) *Towards Completely Recyclable Concrete*, in Proceedings of International Workshop on Rational Design of Concrete Structures under Severe Conditions, 177–186

14. Use of Municipal Waste Island for New Urban Development

Ikuo Towhata

14.1 Introduction

Waste material is generally classified into four categories:

1. Municipal waste produced by citizens and that includes food, paper, and plastics.
2. Industrial waste that consists of oil, chemical materials, and the like, which is sometimes hazardous to human health.
3. Medical waste, which is also known as a dangerous material.
4. Nuclear waste, which is extremely harmful for people and environment.

The present section concerns municipal waste, which has the largest volume among the above-mentioned kinds of waste.

From the environmental and resource-saving viewpoints, the *three-R* principle has been proposed for handling the ever-increasing municipal waste. Three Rs mean **Reduce, Reuse**, and **Recycle**. "Reduce" stands for decreasing the amount of waste as much as possible. "Reuse" means using resources as much as possible; its example being writing on both sides of paper. "Recycle" attempts to use waste materials for other purposes such as converting waste papers to toilet tissue.

The present text concerns the Recycle policy. It is, however, not intended to treat the issue of turning waste material into different materials. The final goal is to use the space occupied by waste landfills for other purposes. This idea may be called recycling in that it deals with the reuse of a spatial resource. To support this idea, the author starts discussion of historical issues, asking why the power of large cities declined in the past. Then an introduction is given about the current situation of municipal waste disposal in Japan, particularly the Tokyo metropolitan area. The third part will describe the mechanical properties of municipal waste as

Y. Fujino, T. Noguchi (eds.) *Stock Management for Sustainable Urban Regeneration*,
© 2009 to the complete printed work by Springer, except as noted. Individual authors
or their assignees retain rights to their respective contributions; reproduced by permission.

obtained through an intense laboratory testing series; here, the advantages and disadvantages of waste as an engineering material are presented. Finally in the fourth section, new applications for waste landfills will be proposed and discussed.

14.2 Mechanism of Decline in the Power of Cities

Human beings started to live together in big groups when agriculture started. Surplus crop harvests in the countryside was sent to a "city" to feed the population there who themselves did not produce food. In other words, city residents somehow collected food from outside, and the way they collected food characterized the nature of a "city." It may well be said that food collection initiated "trading." In a trading market, city and city residents have to bring something that is sold and exchanged for food.

It seems possible to classify cities in accordance with their ways of obtaining food. A political city is ruled by monarchy or a governor and collects tax from its territory. Monetary or otherwise, a tax is a resource for obtaining food from outside. A religious city is a sacred place and earns money by donation. Many pilgrims come there and spend money as well. Thus, the city has sufficient income to buy food from outside. A trading city deals with the flow of commodities and materials. Their handling charge is a good way to earn money. A transportation city is a connection of railways or other means of land and marine transportation. Hotels for travelers, cargo handling, and maintenance of trains and ships earn money. A military city has many officers and soldiers who need food and other services from a local community. For those services, money is paid. A tourism city earns money from tourists who pay money to enjoy sightseeing, accommodation, and other services. A residential city is a place where commuters' family lives and salary money is spent for living expenses.

Thus, any kind of city attracts money in order to sustain residents' lives. If money stops coming in, people have to go to somewhere else and the city declines.

14.2.1 Examples of City Decline

Fujiwara became the first capital city of Japan at the end of the seventh century. The capital was moved to Nara in AD 710. Since government and temples moved to Nara and no income flowed to this former capital, it returned to agricultural use (Fig. 14-1).

Fig. 14-1. Decline of political city

Fig. 14-2. Decline of religious city

Figure 14-2 shows the tomb of the fifth emperor of the Han Dynasty of China. Although the dynasty maintained a small city to take care of the tomb, the city disappeared after the toppling of the dynasty, since there was no more income available.

Turfan was a flourishing trading city on the Silk Road (Fig. 14-3). Many cities along the Silk Road disappeared when ground trading was replaced by sea trading and their income was lost.

No perfect example has been found to illustrate the decline of a tourism city. Recently, however, hot spring resorts around Tokyo have been going downhill because group tours, which used to be a big customer, have decreased in number.

Natural disaster is a unique cause of city decline that is not related to monetary reasons. For example, Yungay of Peru was destroyed by a seismic landslide (Fig. 14-4). Since the city site was then found to be vulnerable to this kind of disaster, it was moved to a safer place. Pompeii in Italy, which was destroyed by a volcanic disaster, is a classic example of the same category.

Kusato-Sengen was an important port town until the fifteenth century. It, however, disappeared later because sedimentary action filled the harbor and its port function was lost. Thus, environmental changes may lead to loss of income and decline of a city.

In summary, history shows that the shortage of money income is the major reason of decline of cities. This is because people come to live in a city in order to get income. Once the money flow stops, people have to move elsewhere. Noteworthy is the fact that beautiful landscape alone does not help people live, unless it brings in sufficient money. In contrast, natural disaster is another mechanism that can affect cities.

Fig. 14-3. Decline of trading city

Fig. 14-4. Kusato-Sengen town

14.2.2 Example of City That Did Not Decline

The former section showed that cities declined when their main function was lost and their income stopped. It is interesting, therefore, to study a city that did not decline despite a loss of function. Nara (Fig. 14-5) became a capital of Japan in AD 710, and governmental institutes came from the former capital (Fujiwara). However, the capital moved in AD 794 to Kyoto (more precisely Nagaoka near Kyoto). Although the political status of Nara was thus lost and tax income stopped, Nara succeeded in maintaining some size for more than 1,000 years after this. The reason for this is because big temples in Nara did not move to Kyoto. Those temples owned huge agricultural lands, which still brought income to Nara. Another reason was that Nara was a center of the local economy.

Kyoto (Fig. 14-6) is another example of a city and maintained its power after the capital function moved to present Tokyo in 1868. After the thirteenth century, Kyoto was no longer the center of political power, and the royal family was kept poor by the Shogun government after the seventeenth century. Thus, the official relocation of the capital in 1868 did not affect the economy of the city. Actually, Kyoto had many big temples and manufacturing industries, both of which brought in monetary income. Efforts of people to modernize the city after Western civilization came were indispensable as well.

Consequently, it is reasonable to state that cities can maintain their power if efforts are made to bring income, even if their previous income mechanism is lost.

Fig. 14-5. Gigantic Todai-ji Temple in Nara

Fig. 14-6. Entrance gate of Imamiya Shrine in Kyoto

14.2.3 Urban Development in UK

The industrial revolution contributed to the development of many cities in England. Those cities encountered difficult times in the second half of the twentieth century due to change of the leading industries. Consequently, city centers lost population, and security became worse. To cope with this, attempts were made to improve the situation by constructing schools, transportation, good apartments, etc. There was, however, no direct plan for economic (monetary) development. The Thatcher administration changed the policy by putting more emphasis on economic development and by advocating privatization and less governmental control.

In Manchester, wealthier residents were invited to live in the city center, and also sporting activities started business. In addition, more cultural events as well as restaurants, theatres, and hotels were developed. Moreover, international investment was facilitated by improving Manchester International Airport. These attempts made Manchester look more attractive than before.

In Liverpool, the old harbor area was converted to a tourist spot with restaurants, high-class hotels, historical spots, and a modern shopping center (Fig. 14-7). It seems evident that such modernization of old city centers was attractive to citizens, and people started to spend more money on these new attractions.

Fig. 14-7. New shopping mall in Liverpool

The author's concern is whether or not the modernization of those British cities succeeded in bringing in external money. This concern comes from the lessons in the previous section that city's survival needs external money to be brought in.

14.3 Use of Municipal Waste Landfill for Urban Development in Future

The preceding sections reveal that the power of cities declines when their means of income are lost. Since the source of income changes continuously in the present world due to changing lifestyles and shifts in leading industries, cities have to be flexible and be prepared for change. One of the disadvantages of Japanese major cities lies in the shortage of open space that can be used to satisfy future demands. Those cities have been expanding in size due to increasing population during the past century, and open space such as green belts was not reserved. This makes it difficult today to obtain space resources without making big efforts in purchasing land.

The present section proposes to use former landfill islands as a space resource. This practice may be called recycling of space of landfill for other good purposes. Many major cities in Japan are located in coastal regions where municipal waste landfills have been constructed on manmade islands in the sea. Some of those islands are located close to the intersection of land, sea, and air transport, and have the advantage of being an international urban and industrial center. More importantly. those landfill spaces are owned by the public sector, so there is no need for time-consuming purchase negotiation.

There are three kinds of municipal waste today: unprocessed waste, incinerated ash, and plastics (incombustible). They come mostly from households and should be strictly distinguished from industrial, medical, and nuclear wastes which are often hazardous to human health.

Although the author is interested in urbanization of former landfills, one significant problem is long-term subsidence (Fig. 14-8), which may damage buildings. It is essential therefore to develop geotechnical engineering that can mitigate and/or terminate the progress of subsidence within a short time.

Types of municipal waste in landfill have been changing. The situation in Fig. 14-8 occurred in an old 1960s-era landfill that was a mixture of all kinds of household unprocessed waste. In more recent times, incineration (burning) has been practiced widely and the resulting ash is dumped in landfills (Fig. 14-9). The ash ground has high rigidity, equivalent to that of sandy ground, and is not vulnerable to any subsidence problem. A problem is posed by the landfill of plastics (Fig. 14-10) which are not incinerated and hence are dumped directly after being cut into small pieces.

The urban area of Tokyo has been expanding continuously towards the sea since the early seventeenth century when the history of this city began.

Fig. 14-8. Damage in building due to subsidence of landfill

Fig. 14-9. Incinerated ash

Fig. 14-10. Dumping site of incombustible waste

While some artificial lands were constructed of soil, others were made of waste. Figure 14-11 demonstrates the most recent landfill construction in Tokyo Harbor. Note that this island is surrounded by an international harbor, an international airport to the left of the figure, and two submarine tunnel connections with downtown Tokyo. Such a good location of a landfill island seems to be found in other cities as well.

Fig. 14-11. Municipal waste island in Tokyo harbor area in operation (©Bureau of Port and Harbor, Tokyo Metropolitan Government)

Fig. 14-12. Stable vertical cliff of waste

14.4 Municipal Waste Mechanics

As shown in Fig. 14-8, long-term subsidence of waste landfill is a significant problem. In this regard, some people believe that municipal waste is a bad material that has very low shear resistance. Figure 14-12, however, shows a cliff that was cut in the landfill of the Göttingen Municipality in Germany and that has been stable for many years. This unexpectedly good performance of waste is produced by inclusions of plastics, paper, and other fibrous materials (Fig. 14-13) that expand laterally and reinforce the entire landfill.

Long-term subsidence was studied by Shimizu et al. (1989) by monitoring the rate of real waste ground in Tokyo. In Fig. 14-14 it seems evident that subsidence lasts for more than 25 years. In contrast, an older landfill from the 1950s has ceased to subside today (Fig. 14-15) and is now used as an urban area without any subsidence problem. Therefore, it seems that 50 years more or less is needed for municipal waste landfill to complete its sub-

Fig. 14-13. Sheets of paper embedded in landfill

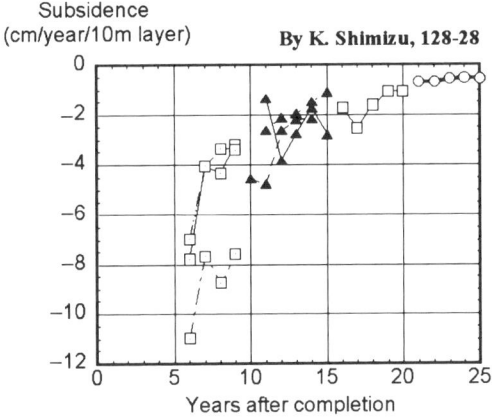

Fig. 14-14. Rate of subsidence of landfill in Tokyo (as per 10 m thickness of waste) (Shimizu et al. (1989))

Fig. 14-15. Urban development in Shiomi area in Tokyo

sidence. Thus, subsidence is the major engineering problem to be studied, and the present investigation attempts to shorten the long duration of subsidence by engineering measures.

Another fear concerns dioxin in the air. Figure 14-16 indicates that the current situation with dioxin in the air is within the limit of regulation, and there is no significant difference between these levels at landfills and those in downtown.

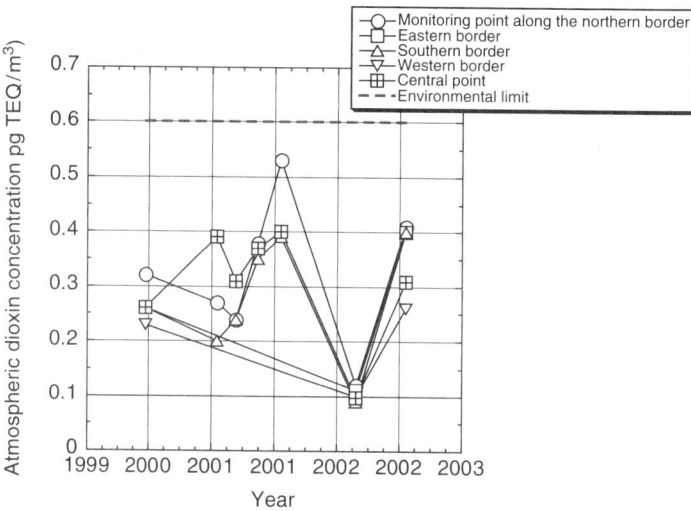

Fig. 14-16. Monitoring dioxin in air in landfill and other places in Tokyo (Data by Tokyo Metropolitan Government)

It should be noted that sea landfills in Tokyo already have had public offices for many decades, and people working there do not have health problems.

Laboratory tests were conducted on municipal waste specimens collected from current landfills in Germany and Tokyo Harbor (Fig. 14-17). The German waste (Fig. 14-18) is a waste treated biologically and includes both combustible and incombustible components.

The Proctor compaction test (Fig. 14-19) is a tool to examine volume reduction of geotechnical materials; repeated impact by a sort of hammer densifies the tested material. Figure 14-19 further illustrates the results of compaction tests in which the dry density (mass of waste dried after compaction) varies with moisture (water) content during compaction. There is a particular moisture content (optimum) at which the obtained dry density achieves the maximum value. This particular moisture content shows that municipal waste has the same nature as discussed in soil mechanics.

Mechanical behavior of municipal waste has been experimentally studied by a triaxial shear device (Fig. 14-20) in which a columnar sample of 25–30 cm in diameter and more than 60 cm in height is sheared by applying horizontal pressure of 40 kPa, for example, and increasing axial (vertical) stress. The vertical stress is increased until the sample achieves large deformation.

Stress-strain behaviors of different wastes are presented in Fig. 14-21. Firstly, the data from German organic waste as shown by solid curves does not show yielding; the stress value increases more or less linearly with strain. When this

Fig. 14-17. Collecting plastic waste in Tokyo landfill

Coarser components

Fig. 14-18. Bio-treated waste imported from Germany

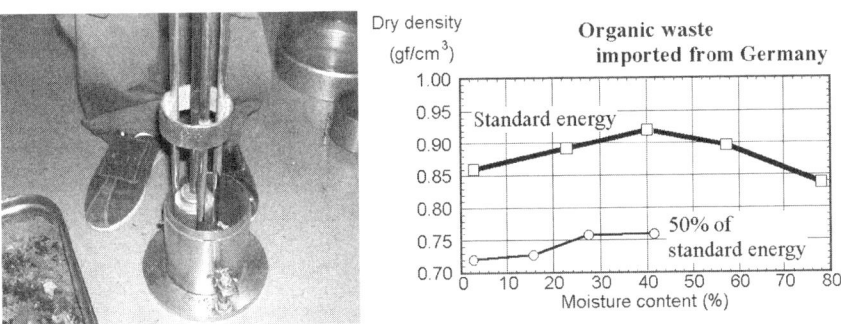

Fig. 14-19. Proctor compaction tests on organic waste

Fig. 14-20. Triaxial shear machine for municipal waste

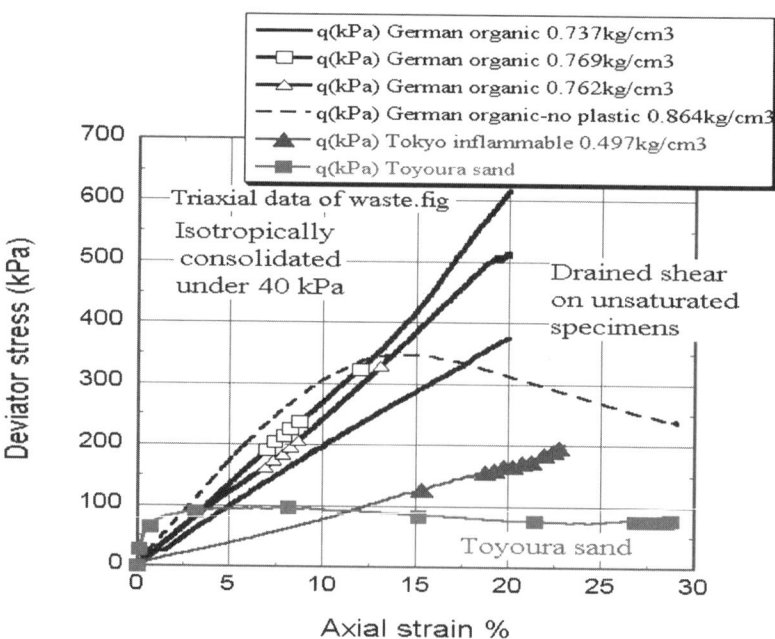

Fig. 14-21. Stress-strain behavior of wastes in triaxial compression tests

behavior is compared with that of sand in the same figure (square symbols), it can be stated that the waste has greater material shear strength than sand. In this regard, it appears that waste is a better building foundation than sand. This slightly strange idea should be considered carefully because the slope of the curves for waste undergoing small strain is smaller than that of sand in the same strain range, indicating that waste ground is much softer than sandy ground. Therefore, the major engineering problem of waste ground lies not in material failure but in large deformation. In the case of a building foundation, the problem is one of subsidence.

It is interesting, as seen in Fig. 14-21, that another stress-strain curve of the German waste (dashed curve) exhibits the peak stress value, followed by a decrease. This result was obtained from a sample from which plastic and fibrous inclusions were removed. As stated before, those inclusions developed lateral tensile resistance and produced the aforementioned linear behavior; after their removal, yielding occurred. This figure further indicates the very soft behavior of plastic waste in Tokyo. Due to such tensile resistance, a linear behavior was obtained again.

Issues involving ground subsidence are now going to be discussed (Fig. 14-22). As was previously shown in Fig. 14-8, subsidence of waste ground is a significant problem that has to be solved prior to urban development. In this regard, volume contraction of plastic waste in a creep manner was studied as an experiment. Figure 14-23 illustrates the finding that volume contraction continues linearly with logarithm of time when the stress level is held constant at 15 kPa.

Preloading is a classical technology in soft-ground engineering in which the mitigation of subsidence of soft clay (mud) ground is the major issue.

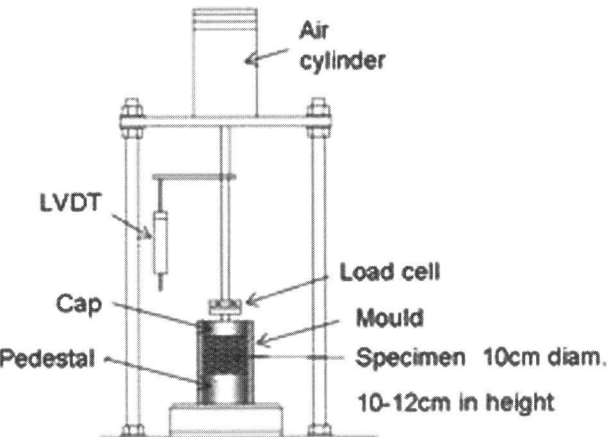

Fig. 14-22. Ground subsidence test on municipal waste

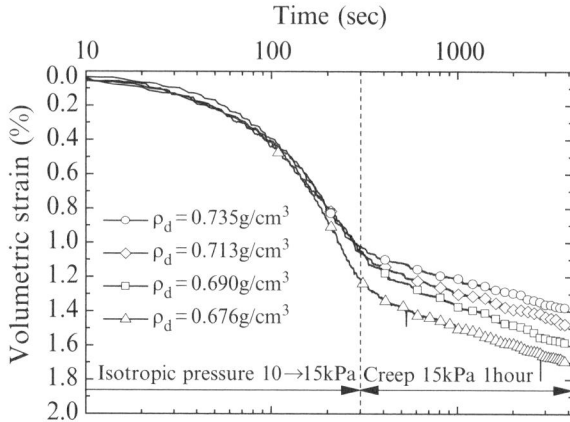

Fig. 14-23. Volume contraction of plastic waste

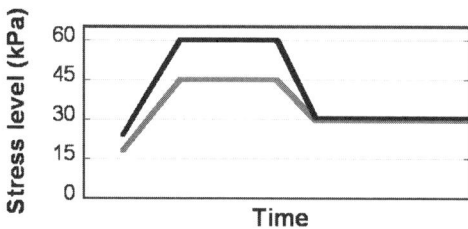

Fig. 14-24. Schematic illustration of preloading technology

In preloading, high pressure is first applied at the ground surface so that the subsoil shrinks very fast. The surface pressure is usually created by placing soil that comes from ground excavation for subway and basement construction. After a sufficient time has passed for the extent of shrinking to be enough, the surface pressure is removed (Fig. 14-24). As suggested by the idea of yield surface in the theory of plasticity, waste becomes stable after stress unloading and the volume contraction is expected to stop.

The variation of volume contraction and subsidence of waste during preloading is illustrated in Fig. 14-24 and Fig. 14-25. Initially, the stress was raised to 45 or 60 kPa, and accordingly the volume decreased quickly with time. The volume further decreased when the stress was maintained constant, which is equivalent with the long-term subsidence of waste ground (Fig. 14-14). When the stress was unloaded to 30 kPa, the volume of the waste stopped decreasing and was held more or less constant. Thus, preloading technology seems to be promising for mitigation of long-term subsidence and it will be possible to shorten the time required for the termination of ground subsidence in waste landfill.

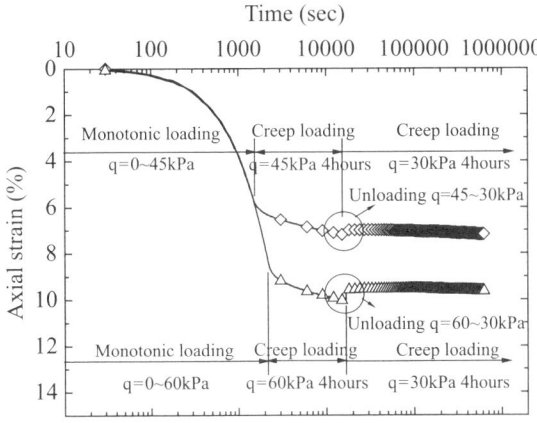

Fig. 14-25. Experimental validation of preloading in mitigation of long-term subsidence (Tested in triaxial device)

Fig. 14-26. Neuss Ski Dome near Köln, Germany

14.5 Proposal for the Use of a Municipal Waste Island for Urban Development

This section addresses the use of former landfill for urban development and future activities. In addition to the example in Tokyo (Fig. 14-15), Fig. 14-26 demonstrates an indoor ski resort that was constructed on a former landfill. This heavy structure is supported by concrete footing blocks embedded in the surface soil that covers the underlying waste (Fig. 14-27). The embedded part of the footing block bears lateral load produced by the inclined shape of the ski building.

Another attempt in Germany was made near Hamburg where a former landfill was improved carefully for residential development (Fig. 14-28).

Fig. 14-27. Big footing foundation upon former landfill

Fig. 14-28. Attempt of residential development at former landfill

To convince clients that no danger existed, some safety certificate was issued. Despite these efforts, nobody purchased this land for fear of, for example, the generation of CH_4 (methane) gas.

Methane gas is produced by anaerobic disintegration of organic materials in waste. This chemical process is dominant when the waste is situated under a ground-water table where insufficient oxygen is available for more rapid disintegration (oxidization). Figure 14-29 shows a chimney with firing mechanism by which methane gas is burned immediately after its discharge from the ground and is converted to carbon dioxide gas (CO_2). This

Fig. 14-29. Burning of methane gas discharge in landfill (Delaware, USA)

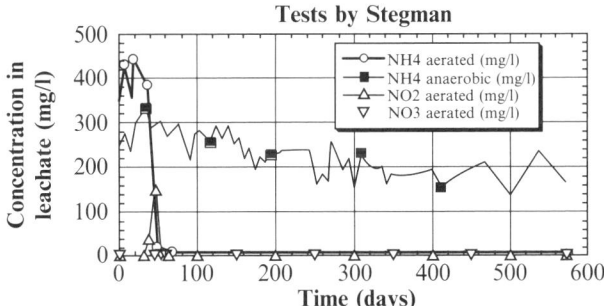

Fig. 14-30. Example of aeration (Data by Stegman, refer to Ritzkowski and Stegman (2005))

procedure seems important because burning can reduce the risk of unexpected explosion of the gas. Moreover, it should be noticed that methane gas has significantly higher green-house effects than carbon dioxide.

To accelerate the anaerobic disintegration of organic materials, environmental geotechnology has been studying aeration (removal of ground water and supplying more air into waste ground) and aerobic procedure, expecting the major gas product changes from methane to carbon dioxide. Experimental data of aeration is presented in Fig. 14-30, which shows that

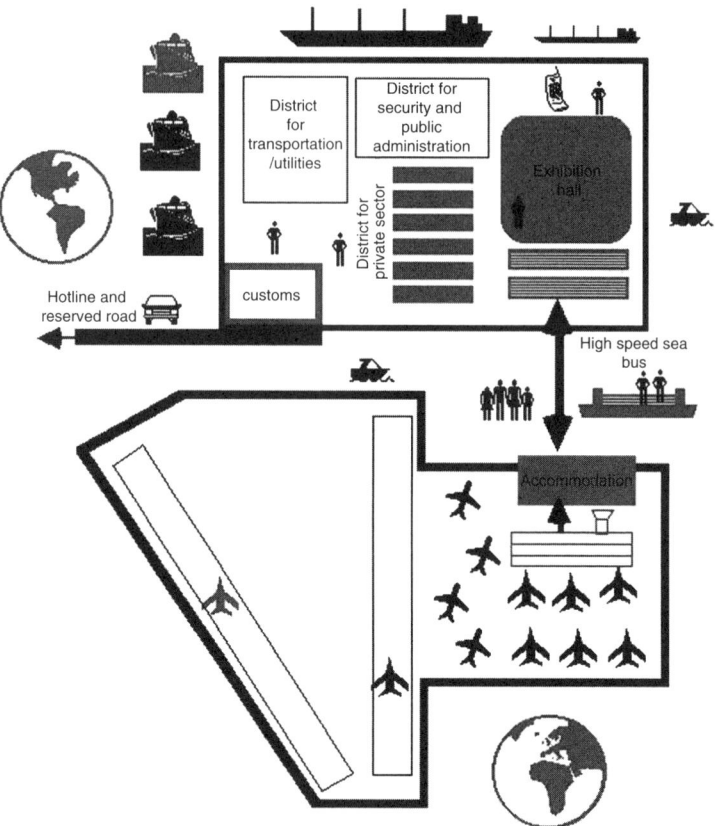

Fig. 14-31. Illustration of international trading center upon former landfill island

aeration induces quick aerobic disintegration and that the produced nitrogen material is changed from NH_3 to NO_x.

With all these in mind, the author proposes to convert a former landfill island near megacity to be an international trading center. The idea illustrated in Fig. 14-31 may be summarized as follows.

1. Since the island is located near an international harbor and airport, it best suits international trading.
2. To produce an unparalleled business environment, this island should be under special laws that allow many business freedoms; for example, exemption of working visas for international business people and quick custom procedures.
3. The island would be an ideal environment to apply different laws. It is easy to control entrance into and exit out of the island.

4. The attached harbor and airport make it easy to quickly deliver goods worldwide that were purchased in the island.

5. An international industry may locate its branch in this island in place of the existing city so that more freedom is obtained in its activity.

Since the success of the above-stated proposal certainly depends on the stability of ground and building foundations, it is important to investigate a suitable type of foundation in addition to ground stabilization by preloading. It is hence supposed that buildings will be supported by friction piles that are floating in waste deposits (Fig. 14-32). End-bearing piles that penetrate through underlying clay barrier layers are not suitable today because this penetration triggers leakage of polluted liquid (leachate) from landfill into the external environment. Since the performance of the friction piles is governed by the magnitude of frictional shear strength between piles and waste, the use of nodular piles with irregular configuration (Fig. 14-33) was investigated. Figure 14-34 illustrates results of model tests in which the variation of friction with vertical displacement of pile was measured. It is evident that planar pile surface (Tests 1 to 4) can develop limited friction, while the irregular nodular surface (Tests 5 to 7) has higher shear resistance (Fig. 14-34).

After considering these findings, finite-element numerical analyses were conducted on a model in Fig. 14-32. The bearing capacity of the foundation was generated as a combined effect of friction piles and the bottom of the basement; this type of foundation is called a piled-raft foundation. As computed in Table 14-1, the combination of preloading on waste and the piled-raft foundation reduce the subsidence to merely 10 percent of what would occur if no mitigation were undertaken.

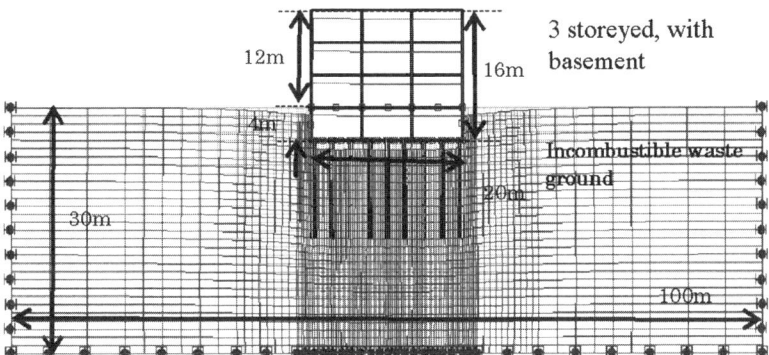

Fig. 14-32. FEM model of three-story building resting upon waste landfill

Fig. 14-33. Nodular piles with irregular surface

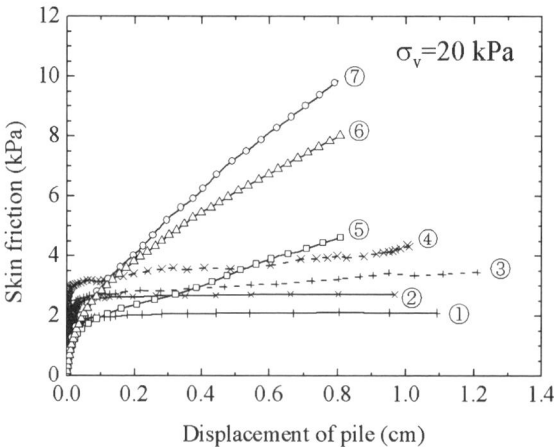

Fig. 14-34. Laboratory model tests on development of friction around pile models

14.6 Conclusions

Historical discussion on the decline of cities reveals that monetary income is the key by which survival or decline of a city is determined. In this regard, the study here intends to convert a huge and public-owned space of municipal

Table 14-1. Reduction of subsidence of building foundation by combined effects of preloading and piled raft

Case	Preloading	Type of foundation	Subsidence (cm)
1	No	Direct	28.6
2	No	Piled raft	9.9
3	Yes	Direct	5.3
4	Yes	Piled raft	3.1

landfill into a space reserved for future urban development to bolster the economic power of a city. This activity may be called recycling of waste space for better purposes. To achieve this goal, it is essential to stabilize in a short time the ground subsidence. Laboratory experiments reveal that preloading technology is useful for this purpose. Moreover, model tests show that the use of nodular and irregular shape of piles decreases drastically the extent of building subsidence. It seems possible therefore to push the idea of an international business center upon municipal waste island near megacities.

References

Shimizu, K., Ebina, S., Saito, S. and Mizukoshi, K. (1989) "Geotechnical Study of Refuse Landfill in Tokyo Port (10)—Subsidence of Refuse Landfill", *Proceedings of 24th Japan National Conference Soil Mechanics and Foundation Engineering*, 1: 149–150.

Ritzkowski, M. and Stegmann, R. (2005) *Mechanisms affecting the leachate quality in the course of landfill in situ aeration*, in Proceedings Sardinia 2005 – 10th International Waste Management and Landfill Symposium, 417–418.